RECENT ADVANCES IN THE CHEMISTRY
OF INSECT CONTROL II

Special Publication No. 79

Recent Advances in the Chemistry of Insect Control II

The Proceedings of the Second International Symposium organised by the Fine Chemical and Medicinals Group of the Industrial Division of the Royal Society of Chemistry and the Pesticides Group of the Society of Chemical Industry

Oxford, 17th—19th July 1989

Edited by
L. Crombie
University of Nottingham

ROYAL
SOCIETY OF
CHEMISTRY

British Library Cataloguing in Publication Data

Recent advances in the chemistry of insect control II.
1. Pests: Insects. Chemical Control
I. Crombie, L.   II. Series
632'.7

ISBN 0-85186-627-1

© The Royal Society of Chemistry 1990

Published by the Royal Society of Chemistry,
Thomas Graham House, Cambridge CB4 4WF

Printed in Great Britain by
Whitstable Litho Printers Ltd., Whitstable, Kent

Editorial Introduction

This book presents, in permanent form, the lectures given at the Second International Symposium, 'Recent Advances in the Chemistry of Insect Control II', held under the auspices of the Royal Society of Chemistry (Industrial Division, Fine Chemicals and Medicinals Group) and the Society of Chemical Industry (Pesticides Group) at St. Catherine's College, Oxford during July 17th. - 19th., 1989. The scientific organisers of the Symposium were Dr. E. McDonald (ICI Agrochemicals), Dr. P.J. Crowley (ICI Agrochemicals), Dr. J.A. Pickett (AFRC Institute of Arable Crops, Rothamsted) and Dr. J.B. Weston (Wellcome Foundation). Administration was in the competent hands of Mrs. Elaine S. Wellingham. All the scientific organisers chaired sessions and were joined in this by Dr. M. Elliott (Lawes Trust, Rothamsted) and Dr. D.A. Evans (ICI Agrochemicals).

The Symposium was built up from four sub-themes: 'Future Trends in Insecticides', 'The Isolation and Synthesis of Insecticidal Natural Products', 'The Invention and Optimisation of Insecticides' and 'Biorational Approaches to Insecticides'. This latter sub-theme reflects the sophisticated and rapidly moving application of modern biology to important problems in target sites. Symposium speakers formed an excellent blend of industry and academia and it is heartening to all who have the problems of agriculture and world food supply in mind to see new approaches to the problem of insecticide resistance, the emergence of new insecticide types, and the sharpening in specificity of action going on within the more familiar groups.

I express my thanks to Dr. N.F. Janes, the Editor of the first volume in this series, and members of the Royal Society of Chemistry Information Services, for their kind help and advice.

September 1989.

Foreword

The symposium, from which this book arose, was the third of
a series, with the first in 1979 also held at St. Catherine's
College and devoted to the pyrethroids which were emerging as a
major class of insecticides. The second meeting in 1984, held
in Queens' College, Cambridge, extended beyond the pyrethroids
to putative insecticides such as the avermectins, active at
other sites in the nervous system, and to compounds influencing
hormonal control of insect development. This second meeting
also marked the retirement, from the AFRC's laboratories at
Rothamsted, of Dr. Michael Elliott, CBE, FRS, who played such a
key role in developing the pyrethroid insecticides from the
original natural product lead compounds in pyrethrum extract.

Without efficient crop protection, there is no doubt that
we would not have sufficient food or natural fibre. Indeed, in
the developing countries of the world there is still an enormous
need for increased crop protection. In the case of insect-pest
control, considerable efforts must be made to ensure that there
is no danger to man since the biochemical targets in insects can
be closely related to our own. This requirement has meant that
there has been, and indeed still is, a major research effort on
insect control. This has paid off admirably by producing a wide
range of insecticides safe to man and with minimum impact on the

environment. In the 1920s, a recommended control measure against various noxious insects was "Kansas Bait", comprising a mixture of Paris green, molasses, bran and lemons. Paris green, the arsenic-containing pigment copper acetoarsenite, was the toxicant, and the use of such a generally toxic material well into the century serves to remind us of just how far we have come in producing selective insect control agents. This meeting was held 50 years after the first use of DDT as an insecticide, which itself represented a tremendous leap forward in being a very active insecticide with virtually no mammalian toxicity. Of course, it was soon found that DDT persisted for too long in the environment, but this has been allowed to eclipse the immense value that this compound has had, for example in controlling human disease vectors. We then saw the development of the less persistent organophosphorus and carbamate insecticides and then the pyrethroids, all demonstrating the ability of the many facets of agricultural research to produce chemical compounds tuned to the changing needs of agriculture, health and environmental protection. Now we see further commercial compounds attacking a range of targets in the insect. With the advent of genetic engineering, we also see innovative attempts to deliver insecticidal gene products to pests.

In spite of the rapid commercial developments in agents toxic to insects, there has been a limited commercial impact by behaviourally active materials, exemplified earlier as the insect attractant components of the lemons in Kansas Bait. However, in order to provide even more benign methods of insect control and also to provide control measures capable of dealing with insects resistant to current agents, we must in the future give more attention to the subject. It was therefore appropriate that at this symposium we considered such behaviourally-active chemicals and the subject of insecticide resistance.

Although we must strive to reduce still further any possible hazard from insect control, we must not accept the view that current insect control presents a real problem especially when set against the value of such measures and the hazards from many other human activities. However, it will only be by publicising our past achievements and by showing how we are working for the future that advances in the chemistry of insect control will be seen in their true light.

J. A. Pickett

Contents

Today's Research for Tomorrow's Markets or: How to Hit a Moving Target

Ch. v. Szczepanski

AGROCHEMICAL RESEARCH, SCHERING AG, D-1000 BERLIN 65, FRG

1 INTRODUCTION

This Symposium is providing various answers on the future of insect control in the areas of natural products, synthetic insecticides and biorational approaches. Why are we, why is industry then worrying about tomorrow's markets, why are we not daring to follow the easy going slogan 'Don't worry, be inventive?'

Targets move on the long path from compound discovery to market. Because of the expensive and time consuming lead finding and compound development process, the vital question within industrial research and development must be: what are tomorrow's needs, where will today's markets move to within the time span of product innovation?

We will look at this moving scene beginning with a view on today's insecticide markets to be described in terms of crops, pests and compounds currently used. The moving targets issue can best be addressed by describing factors promoting change and I will illustrate the dynamics of this change with examples drawn from our own experience.

Focussing today's research on tomorrow's markets must mean relating the inventive process to just those factors of change. Again, I will be specific about our approaches by using examples close to our work. My conclusions at the end have again question-marks, as there will be no certainty about the future, but conviction and hope.

2 WHY TARGETS MOVE

Table 1 is aimed at explaining the importance of the moving targets issue. Compound discovery and product development can be - somewhat artificially-split into 4 stages: first synthesis and screening; repreparations and confirming activity under field conditions; chemical process development, field trials on a broader scale, environmental and toxicity testing; finally the completion of all product development, safety and efficacy tests including registration by local government authorities. The minimum time required 'from bench to farm' is around 8 years. The second column of this Table shows the average number of candidate compounds which must enter each stage in order to forward one successful compound to the next. Our experience tells us that 150 compounds have to be synthesised and tested to finally obtain one candidate for first field experiments. At the end, we arrive - through all the 4 stages - at a success rate of 1 commercial compound out of 20,000 compounds synthesised.

Table 1: Time scale, success rate and cost of agrochemical compound research and development				
Stage	Minimum duration / years	Success rate / 1 : n	Costs / '000 DM	
			per successful compound	incl. unsuccessful compounds
1st synthesis / tests	1	150	6	135,000
Repreparations and 1st field tests	2	20	22	3,300
Process development, further field trials, safety evaluation	2	5	2,700	20,250
Complete development to registration	3	1,5	17,100	25,650
From bench to farm	≥ 8	20,000	20,000	184,000

The third column shows that one successful compound costs roughly 20 Mio DM and that the later steps - mainly toxicity tests and environmental studies - are most expensive. The picture changes dramatically if all unsuccessful compounds are included. We then arrive at an average cost per registered product of 180 Mio DM (60 Mio £) with the majority of the costs being linked to the selection of the very few compounds for further development. Thus, the efficiency of compound discovery is highly depending upon our ability to select the most promising compounds at an early stage.

In short, today's research has to look well ahead because 'tomorrow' lies at least 8 years in the future. In view of the cost incurred, we have to concentrate on the major crop/insect targets.

3 INSECTICIDES TODAY

Today, the most important single crops with regard to insecticide usage are cotton, rice and maize; taking related crops together, vegetables and fruit also represent major target crops.

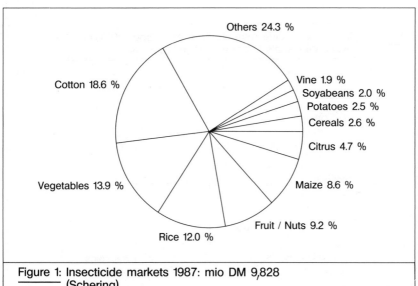

Figure 1: Insecticide markets 1987: mio DM 9,828
———— (Schering)

Target insects are shown in Figure 2: <u>Heliothis</u> in cotton and <u>Diabrotica</u> in corn are the most important single targets; spider mites (mainly on fruit and cotton), aphids (<u>e.g.</u> on sugar beet and cereals), stemborers on rice and maize and hoppers on rice represent further major insect target groups.

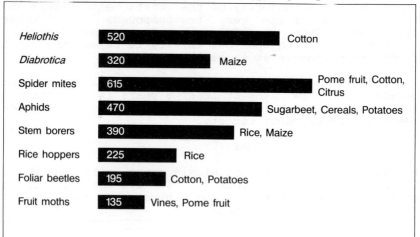

Figure 2: Major insecticide markets and main crops 1987/mio DM (National Panels / Schering, Distributor level)

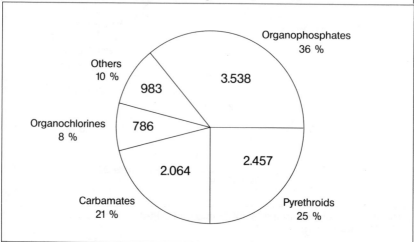

Figure 3: Insecticide markets by compound class 1987/mio DM (WM / Schering, End user level)

A view on the insecticide market by compound class reveals a rather conservative picture with organophosphates still dominating, followed by pyrethroids and carbamates.

4 FACTORS PROMOTING CHANGE

Table 2 summarises a number of interdependent factors promoting change in today's insecticide markets.

Table 2: Factors promoting change in the use of insecticides	
Crops	choice of crop acreage, new (resistant) varieties, new crops
Pests	fluctuating populations, resistance, new pests
Insecticides	spectrum, level of activity, mode of action
Safety	consumer, farmer, environment
Economics	R & D costs, cost of product, farmers income
Public awareness	maximum crop quality, no risk
Legislation	product restrictions, government influence on planted acreage, subsidies

Crops do and will change, both in quantitative importance and in quality. The area grown is certainly strongly determined by economic factors which are in turn influenced by government actions such as set aside programmes or subsidies. There is much discussion about alternative crops for non-food use. Personally, I would not expect this to become an important factor outside political discussions. At a recent meeting of experts on renewable resources[1], all non-food applications together were estimated as being equivalent to the increase in productivity achieved in the European Community within 3 to 4 years. The breeding of insect resistant crop varieties will however reduce insecticide usage in some crops. This influence will increase through the application of recombinant DNA technology leading to the introduction of transgenic crop plants.

Pest populations will continue to fluctuate following climatic changes and cropping practice. Resistance to insecticides has a major influence because existing products are based on only a few active compound classes with essentially only two different modes of action: the organophosphates and carbamates affecting the cholinergic synapse by inhibition of acetylcholine esterase and the pyrethroids acting on the voltage sensitive sodium channel. Important pests are appearing in areas where they did not occur before. This will create the need for insect control programmes based on products with new qualities in terms of activity and mode of action. Safety - to the consumer, to the farmer and to the environment - is an area of growing awareness to the public and to governments as well. This is the obvious concern of industry too. Clearly, solutions to problems also have to be economically acceptable. It is difficult to fulfil additional requirements under conditions of economic stringency such as rising R & D and product costs and falling farm income in many important countries.

Let me illustrate this changing scene giving a few examples.

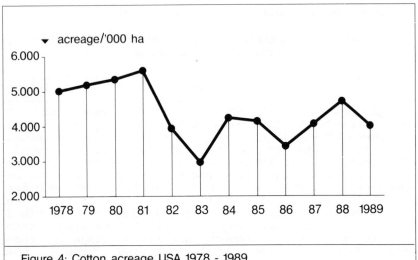

Figure 4: Cotton acreage USA 1978 - 1989
         (FAO)

Cotton acreage in the US was negatively influenced by the introduction of the 'Payment in Kind' (PIK)-program in 1982: cotton acreage dropped by almost 50 % within two years from 1981 to 1983.

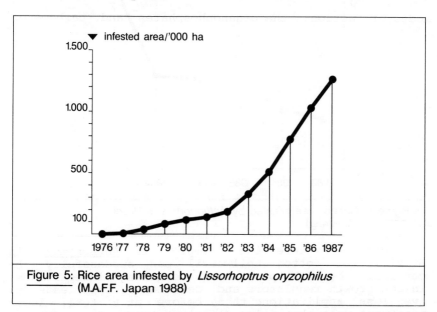

Figure 5: Rice area infested by *Lissorhoptrus oryzophilus* (M.A.F.F. Japan 1988)

The rice water weevil <u>Lissorhoptrus oryzophilus</u> was first discovered in Japan 1976. The introduction of this species - most probably from California[2] - has added over 80 mio DM to insecticide rice treatment costs in Japan without solving the problem: there is a need for new active compounds combining larvicidal and adulticidal efficacy with safety to the crop, to fish and to man.

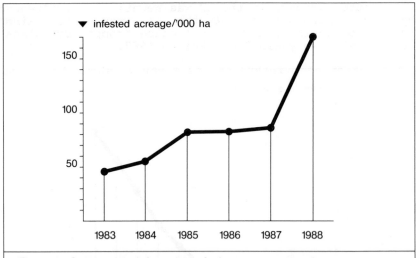

Figure 6: Cotton area infested by *Anthonomus* in Brazil
(Schering)

Another interesting example is <u>Anthonomus</u> in Brazil. The cotton bollweevil can successfully be controlled by IPM programmes combining insecticides, insect growth regulators and defoliants. This leads to even less applications than before at a total lower treatment cost, but using different types of products.

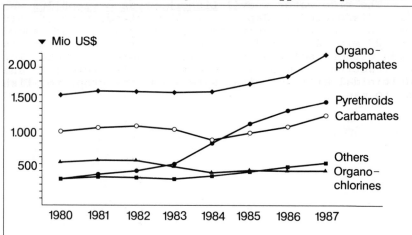

Figure 7: Worldwide use of insecticides 1980 - 1987
(WM, End user level)

It is remarkable indeed, that in contrast to this changing problem scene, 'markets' in terms of chemicals applied did not change much.

In fact, the only major success was the introduction of pyrethroids as clearly shown from the only rising curve in this Figure. Of course, this is the overall picture and the substantial contribution of some new compounds to smaller market segments cannot be seen. Without any doubt, there is room for innovation - but with increasing stringent requirements.

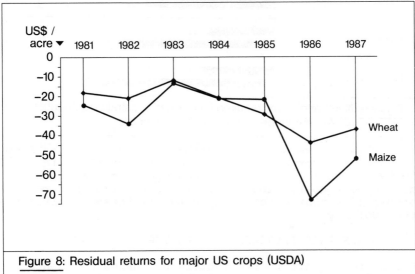

Figure 8: Residual returns for major US crops (USDA)

Farm income is decreasing in many agricultural advanced regions of the world. Figure 8 shows residual returns per acre of wheat and maize in the US taken from USDA sources. Residual return means receipts (excluding government payments) minus expenses and full ownership costs: the economic squeeze on US agriculture is evident. Another dominant factor is increasing environmental concern.

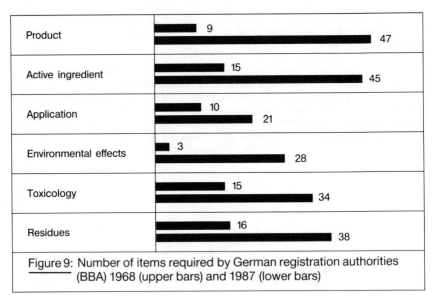

Figure 9: Number of items required by German registration authorities
(BBA) 1968 (upper bars) and 1987 (lower bars)

This Figure shows the number of registration items required by German registration authorities in 1987 compared to 1968: the consequences are higher development cost, longer development time and reduced probability that a compound meets all these requirements successfully.

5 RESEARCH APPROACHES TO FACTORS OF CHANGE

Let us now revert to the factors of change as mentioned again in Table 3. The aim of this chapter is to link the changing scene to research approaches which might help us in providing solutions for the insect problems of the future: today's research for tomorrow's markets. Again, we will first look at the general picture and then illustrate it with examples.

Natural defence mechanisms of plants against insects are widely spread but only partially known and even less well understood. Research into those natural defence systems offers a real chance to build on nature's experience accumulated through evolution. If proteins are involved, plant molecular genetics offer the tools for organ specific expression to be induced by insect attack. The availability of proteins within plants, at the right time and location offers a precision of targeting far beyond that classical insecticide treatments can achieve.

Table 3: Research approaches in response to factors promoting change	
Crops	elucidation of natural defence mechanisms, defence proteins to be expressed, organ- and insect-specific expression systems
Pests	target screening, tests on resistant species
Insecticides	compound discovery based on molecular / cellular mode of action
Safety, public awareness and legislation	activity and safety of equal importance in compound research and development
Economics	minimum structure for activity required, efficiency in compound discovery

We should not neglect, however, that the potential advantages of engineered crop varieties may be counterbalanced by uncertainties which still need clarification. I am referring to the rate at which resistance will arise, to unresolved regulatory questions and to the availability of patent protection.

The use of biologicals as insecticides should also be mentioned in this context. Up to now, bacteria, fungi, virus, protozoa and nematodes were of minor practical importance due to their restricted spectrum of activity, inconsistent performance and high production costs. This may change in the future, because genetic engineering of microorganisms, a relatively easy exercise, may create biologicals with broader and more reliable activity.

New compound research has not only to know about its insect targets but also to have them available for testing from the very first stage. Resistant species have also to be included. In view of the restricted number of modes of action available and the real threat of cross resistance to different chemical structures having the same mode of action (take OPs and carbamates as classical examples), there is a strong need for compound discovery based on new modes of action.

Modern new compound research offers more opportunities for a systematic approach to lead finding based on biochemistry and physiology and to direct and monitor structure optimisation by adding these new methods to the classical in vivo tests.

Safety aspects are now of equal importance as activity within compound research and development will be considered much earlier than previously. Economics within new compound research ultimately means optimising chemical structure requirements for biological activity in order to arrive at the simplest chemical structure required to obtain the desired biological activity needed. Molecular modelling, both on the insecticide itself and on the receptor target site offers opportunities which we are just beginning to exploit. Let me now refer to some of these research approaches in more detail.

Table 4 gives some examples of proteins with proven or hypothetical potential as defence proteins.

Table 4: Some examples of plant defence proteins

- Bt - toxin

- Protease inhibitors

- Lectins

- Chitinase

- Enzymes

The case of Bt-toxin is as well known as the picture of Plant Genetic Systems' transgenic tobacco plants hardly damaged by the tobacco hornworm <u>Manduca sexta</u>[3]. A similar case, the engineering of the cowpea trypsin inhibitor into tobacco was published by Boulter[4] together with scientists from the Plant Breeding Institute. Lectins, carbohydrate binding proteins interacting with the larval gut epithelium, were suggested by Boulter[5] within the same context. The chitin layer of the larval gut could explain the effect of chitinase applied by injection. Enzymes which might have an effect either themselves or by generating the active defence molecule through synthetic or hydrolytic action would be another approach, superbly exemplified by B. Hammock and his work on juvenile hormone esterase reported at this Symposium. The case of protease inhibitors may be taken further by adding to the picture the systemic activation of a defence protein gene by insect attack. As shown by Willmitzer in Berlin and Ryan at Pullman/University of Washington[6] a proteinase inhibitor inducing factor is released at the wound site of the plant and transported to non wounded stems and leaves where it is organ-specifically expressed.

Based on the experience of finding activity in model test organisms which could not be confirmed in the real targets later, we had to learn some rather expensive lessons on the importance of target screening.

Table 5: Some examples of target screening (Schering 1980 / 1989)

Pest	Test species used	
	1980	1989
Heliothis / cotton	*Plutella xylostella* *Spodoptera littoralis*	*Heliothis virescens*
Diabrotica / maize	*Epilachna varivestis*	*Diabrotica undecimpunctata*
Nilaparvata lugens / rice	*Aphis fabae*	*Nilaparvata lugens*

This is clearly illustrated by looking at some of the
test species used in our primary insecticide screening,
as it is now and as it was 10 years ago. The shift from
screening against model insects to using the target
insects from the first stage on is obvious.

Environmental and consumer safety now receives
earlier attention in new compound research. While the
lead finding part of compound discovery is still largely
directed by the search for activity through a desirable
mode of action, the optimisation of a lead structure is
governed by consideration of both activity and safety.

Table 6: Environmental parameters to be considered in early compound research (soil insecticide)	
Synthesis planning	water solubility, log p, vapour pressure
1st synthesis / screening	soil half life, leaching
Repreparation and selection for field testing	soil adsorption, microbial degradation, light / hydrolytic stability. Acute / subchronic toxicity

Physicochemical parameters linked to environmental
properties are part of the thought process from
synthesis planning onwards and appropriate tests are
available at an early stage. There is also a need to do
preliminary toxicity testing at earlier stages, mainly
in those cases where the nature of the toxic effects can
be clearly defined.

Last not least, the question of economics arises.

Table 7: The economics of compound discovery – some aspects
to consider

- Structure / activity correlation through CAMM, QSAR

- Stereospecific synthesis, biotransformations

- Relevant biological tests (incl. environmental effects and toxicity), early enough

- Basic approaches based on desirable mode of action

- Formulation / application

- Good / quick reporting of results

Economics as applied to the discovery of new compounds means reaching the biological target with the simplest molecule and as efficiently as possible. Computer modelling certainly is a tool for discovering the common core of different active molecules sharing the same mode of action. Stereospecific synthesis and biotransformations will play an increasing role in producing chiral insecticides at reasonable cost. The need to use relevant tests including the desired activity as well as mammalian toxicity and environmental properties at an early stage has already been mentioned. Formulation of the active material and appropriate application techniques should be added as further essentials for a new agrochemical product. It is obvious that communications, e.q. the rapid feedback of results, is a key success factor in research organisations characterised by close cooperation between representatives of various disciplines.

6 THE FUTURE: CHANCES THROUGH CHANGE?

In summarising the 'moving targets' issue and pinpointing the role of research within this context, the essence would be: chances through change!

While we cannot predict the future, we can clearly recognise some factors which will influence tomorrow's insecticide markets. Some lead us to predict that crops may need fewer insecticide treatments as we will be able to strengthen their own potential for defence. Other factors promoting change call for improved product qualities related to activity and safety. We will be able to address these more successfully in the future by a better targeting of our approaches in chemical synthesis, biochemical, biological and environmental testing. Last not least, we need well trained and enthusiastic scientists, flexibility within our organisations and more cooperation with basic research institutes outside industry.

I am happy to acknowledge the contributions of my colleagues at Schering: H. Gabriel, D. Giles, H. v. Keyserlingk, G. Kuschlanski, H. Neh and H. von Stillfried.

REFERENCES

1. W. Henrichsmeyer at Expertenkolloquium "Nachwachsende Rohstoffe", Wissenschaftszentrum Bonn, 14. - 15. October 1986

2. M. Matsui, JARQ, 1987, 20, 166

3. M. Vaeck, A. Reynaerts, H. Höfte, S. Jansens, M. De Beuckeleer, C. Dean, M. Zabeau, M. Van Montague and J. Leemans, Nature (London), 1987, 328, 33

4. V.A. Hilder, A.M.R. Gatehouse, S.E. Sheerman, R.F. Barker and D. Boulter, Nature (London), 1987, 300, 160

5. D. Boulter, Outlook on Agric., 1989, 18, 2

6. M. Keil, J.J. Sànchez-Serrano and L. Willmitzer, EMBO J., 1989, 8, 1323

Biochemical and Molecular Biological Approaches to Insecticide Resistance Research

Alan L. Devonshire

AFRC INSTITUTE OF ARABLE CROPS RESEARCH, INSECTICIDES AND FUNGICIDES
DEPARTMENT, ROTHAMSTED EXPERIMENTAL STATION, HARPENDEN, HERTS. AL5 2JQ,
UK

1 INTRODUCTION

The scientific approaches to studying insecticide
resistance are evolving just as are the insects
themselves. In its early days, resistance was
characterised solely by the response of insects to
insecticides in the field or in laboratory bioassays.
Whilst bioassays must remain at the centre of all
resistance studies, they are now complemented by other
methods that can throw light on the underlying
biochemical and genetic changes responsible, and on how
these changes develop within insect populations:-

The use of synergists in bioassays can give <u>indications</u>
of the biochemical mechanisms involved.

Formal genetic studies establish inheritance patterns
and can isolate resistance genes for detailed
toxicological analysis.

Population genetics and modelling describe and predict
the build up of resistance.

Studies of insecticide metabolism can identify the type
of enzyme(s) involved in resistance.

Protein purification and characterisation can establish
whether insecticide degrading enzymes change
qualitatively or quantitatively in resistant insects.

Electrophysiological and enzyme kinetic studies describe
changes in the interaction between insecticides and
their targets.

Molecular biological techniques elucidate the changes in
DNA underlying all the above observations.

Biochemical, immunological and molecular biological
studies can provide accurate and sensitive methods for
monitoring resistance genes or their protein products in
insect populations.

It is only by exploiting all these approaches to
understand the factors influencing the build up of
resistance that we can hope to slow down this almost
inevitable evolutionary consequence of man's activities.

2. BIOCHEMISTRY

The role of biochemistry is to explain how resistant
insects avoid being killed. This can help in devising
strategies of insecticide use to limit their build up
and spread in populations.[1] For example, it is
important to know whether a particular instance of cross
resistance results from one resistance mechanism or
several; if only one, there is nothing to be gained by
mixing or alternating those insecticides affected[2]
whereas this approach could be exploited when multiple
genes are involved. Similarly, when the degree of
resistance conferred by a single biochemical mechanism
varies towards different insecticides, then the use of
those chemicals least affected is likely to slow the
build up of that resistance gene in the population.[3] A
spin-off from the fundamental study of resistance
mechanisms has been the development of biochemical[4] and
immunological[5] monitoring techniques, which can be used
to measure resistance gene frequency in populations and
how it changes in response to different treatment
regimes.[3] Such information is helpful to population
geneticists modelling the development of resistance.
Finally, in view of the relatively small number of
targets for insecticides and the few mechanisms for
their detoxication, a good biochemical insight of their
interactions in susceptible and resistant insects should
help guide the development of better insecticides.[1]

3. MOLECULAR BIOLOGY

Whilst biochemistry is a well established component of
resistance research, molecular biology is only beginning
to have an impact. Besides identifying the DNA changes
responsible for resistance, it will also provide probes
for monitoring resistance genes directly in insect
populations.[6,7]

Insecticide degrading enzymes

The molecular genetic basis of resistance has been established for only two insects, the aphid *Myzus persicae*[8] and the mosquito *Culex quinquefasciatus*.[9] Both rely on gene amplification, i.e. each resistant individual carries multiple copies of a DNA sequence, to produce large amounts of esterase which, in the case of the aphid, has been shown to detoxify a broad range of insecticidal esters by both hydrolysis[2] and 'sequestering'[10] when its catalytic centre becomes acylated. In aphids, the amount of esterase produced is not simply a function of the number of esterase genes present; transcription of the amplified DNA into RNA can be regulated, apparently involving changes in the degree of cytosine methylation in or around the esterase genes,[11] and this can result in the spontaneous loss of resistance within non-selected aphid clones.[12] Gene amplification may be a common genetic phenomenon responsible for other examples of increased esterase activity in resistant insects, as well as of other enzymes such as DDT dehydrochlorinase and glutathione transferases,[13] although there is not yet any direct evidence. Indeed it is rarely known whether these increases in enzyme activity arise from a qualitative or quantitative change in the proteins. However, the momentum of molecular biology is such that many of these uncertainties are likely to be resolved within a few years.

The recent cloning of a cytochrome P-450 cDNA from insecticide resistant houseflies[14] is a significant step forward in understanding the regulation of this gene family and its role in insecticide resistance. Published work relates only the gene in resistant houseflies, but when used for comparative studies with susceptible insects, the probe should establish whether increased monooxygenase activity causes resistance as a result of mutations increasing expression of the cytochrome P-450 gene(s), or of others that change the catalytic properties of the P-450 protein(s). It will also throw light on the multiplicity of enzyme forms, which in turn should help explain the substrate (insecticide) specificities of monooxygenase systems. Similarly, the cloning of glutathione transferase genes from insects, as with the mammalian genes,[15] should clarify the role of multiple enzyme forms and the qualitative and quantitative changes that occur when resistant insects acquire increased transferase activity.[13]

Insecticide target proteins

Resistance often arises from changes in the target
proteins for insecticides.[13] This qualitative variation
is well established for acetylcholinesterase which can
exist in multiple allelic forms each with a characteristic spectrum of sensitivity to organophosphorus and
carbamate insecticides.[16] Similarly, resistance to
pyrethroids and DDT can be caused by changes in nerve
sensitivity,[13] identified electrophysiologically and
again probably involving modification of their target
sodium channels.

Acetylcholinesterase[17] and sodium channel[18] genes
have been cloned from several organisms including
Drosophila, and these should allow the cloning of
corresponding sequences from pest species, either by
using the cloned genes themselves as heterologous probes
or by synthesising appropriate oligonucleotides as
probes. The isolation and characterisation of genes for
the normal and insensitive target proteins will pinpoint
those regions critical for their interaction with
insecticides. This should be especially informative for
the series of acetylcholinesterase forms already
identified in the housefly. It will then be possible to
analyse these variable regions in more detail by site
directed mutagenesis which allows the controlled
modification of amino acid sequence when the genes are
expressed, for example in *Xenopus* oocytes. The
consequences of such modifications can be assessed by
functional assays involving enzyme kinetics for
acetylcholinesterase or electrophysiology for sodium
channels incorporated into the oocyte membrane.[19] This
will provide detailed structural information on the two
major targets for insecticides and so contribute to the
rational design of insecticides effective against the
insensitive forms.

Engineering resistant insects

Cloned genes will not only provide information
directly on the proteins and consequent processes that
mediate resistance; they will also be used to transform
living insects in order both to determine the functional
consequences of the changes observed, and to confer
resistance on the recipient. So far the techniques for
insect transformation are developed only for Drosophila
where a naturally-occurring transposable element
(P-element) has been used as a vector to introduce
foreign genes into the germline so leading to stable

inheritance.[20] Attempts to transform other insects using the P-element as vector have met with only limited success,[21,22] and even then it was not clear whether transformation was P-element mediated or simply arose by recombination of the introduced DNA with homologous sequences in the recipient's genome. However, once such techniques become widely established, and resistance genes and their controlling elements are cloned and characterized, there will be opportunities for engineering resistance into beneficial insects such as honeybees and insect parasitoids and predators.

In addition to this direct potential contribution to integrated pest management programmes, conditionally-expressed resistance genes could be introduced into those species amenable to the sterile male insect release technique.[23] This would have a major impact on its economics by allowing the removal of superfluous females at an early stage of the rearing procedure. Besides the large savings in rearing costs, it would remove the opportunity for preferential mating between released insects and also avoid the damage that can be caused by sterile females during feeding or in their abortive attempts to oviposit.[24] Whilst the possible rewards of such developments are great, the potential environmental consequences will have to be very carefully assessed before releasing transgenic insects.

REFERENCES

1. 'Pesticide Resistance: Strategies and Tactics for Management', (edited by National Research Council Committee on Strategies for the Management of Pesticide Resistant Pest Populations) National Academy Press, Washington D.C., 1986, Chapter 2, p. 45.
2. A.L. Devonshire and G.D. Moores, Pestic. Biochem. Physiol., 1982, 18, 235.
3. R.H. ffrench-Constant, A.L. Devonshire and S.J. Clark, Bull. entomol. Res., 1987, 77, 227.
4. G.D. Moores, A.L. Devonshire and I. Denholm, Bull. entomol. Res., 1988, 78, 537.
5. A.L. Devonshire, G.D. Moores and R.H. ffrench-Constant, Bull. entomol. Res., 1986, 76, 97.
6. T.M. Brown and W.G. Brogdon, Ann. Rev. Entomol., 1987, 32, 145.
7. L.M. Field, A.L. Devonshire, R.H. ffrench-Constant and B.G. Forde, Pestic. Biochem. Physiol., 1989, 34, in press.

8. L.M. Field, A.L. Devonshire and B.G. Forde, Biochem. J., 1988, 251, 309.
9. C. Mouchès, N. Pasteur, J.B. Bergé, O. Hyrien, M. Raymond, B. Robert de Saint Vincent, M. de Silvestri and G.P. Georghiou, Science, 1986, 233, 778.
10. A.L. Devonshire and G.D. Moores, 'Enzymes Hydrolysing Organophosphorus Compounds' (editors E. Reiner, W.N. Aldridge and F.C.G. Hoskin), Ellis Horwood, Chichester, 1989, Chapter 16, p. 180.
11. L.M. Field, A.L. Devonshire, R.H. ffrench-Constant and B.G. Forde, FEBS Lett., 1989, 243, 323.
12. R.H. ffrench-Constant, A.L. Devonshire and R.P. White, Pestic. Biochem. Physiol., 1988, 30, 1.
13. F.J. Oppenoorth, 'Comprehensive Insect Physiology, Biochemistry and Pharmacology' (editors G.S. Kerkut and L.I. Gilbert), Pergamon, Oxford, 1985, Chapter 19, p. 731.
14. R. Feyereisen, J.F. Koener, D.E. Farnsworth and D.W. Nebert, Proc. Natl. Acad. Sci. USA, 1989, 86, 1465.
15. B. Mannervik and U.H. Danielson, CRC Crit. Rev. Biochem., 1988, 23, 283.
16. G.D. Moores, A.L. Devonshire and I. Denholm, Bull. entomol. Res. 1988, 78, 537.
17. L.M.C. Hall and P. Spierer, EMBO J., 1986, 5, 2949.
18. M. Ramaswami and M.A. Tanouye, Proc. Natl. Acad. Sci. USA, 1989, 86, 2079.
19. H. Suzuki, S. Beckh, H. Kubo, N. Yahagi, H. Ishida, T. Kayano, M. Noda and S. Numa, FEBS Lett., 1988, 228, 195.
20. G.M. Rubin, Trends Neurosci., 1985 (June), 231.
21. L.H. Miller, R.K. Sakai, P. Romans, R.W. Gwadz, P. Kantoff and H.G. Coon, Science, 1987, 237, 779.
22. A.C. Morris, P. Eggleston and J.M. Crampton, Med. Vet. Entomol., 1989, 3, 1.
23. A.S. Robinson, C. Savakis and C. Louis, 'Modern Insect Control: Nuclear Techniques and Biotechnology', IAEA, Vienna, 1988, p. 241.
24. C.F. Curtis, Biol. J. Linn. Soc., 1985, 26, 359.

Natural Models for the Design of Insect Control Compounds: The Mammeins

L. Crombie
DEPARTMENT OF CHEMISTRY, THE UNIVERSITY OF NOTTINGHAM, NOTTINGHAM NG7 2RD, UK

Although ideally one would like to see new insect control agents emerge logically from increasing biochemical understanding, screening of synthetic chemicals to obtain leads for furthur development of insecticidal activity is still part of the routine work of many laboratories. Natural man since distant times has, using plants as his chemicals source , and in a disorganised way, been doing much the same thing with the aim of alleviating insect parasitism and protecting his food supplies. So we have traditions from many cultures about which plants possess insecticidal compounds. In more recent times the screening of plants has been extended into areas much less accessible to earlier man, including micro-organisms and marine plants and animals, but the exploitation of the more traditional areas of higher plant insecticides is still far from complete and forms a valuable basis for new insecticide development.

Verification of traditional claims, isolation of pure active components, and determination of their structure and stereochemistry, remain essential to bring a prospect to the starting line. Total synthesis and structure - activity relations follow naturally if the system has adequate properties and activities. With large and complex structures, except inasmuch as total synthesis may reveal segments which hold activity, total synthesis may be more of an academic challenge than a practical base for commercial development: it is hard to compete with the economics of micro-organisms in making complex structures. In such cases a programme of reconstructive synthesis - selective degradation and synthetic replacement of molecular segments - may be a profitable way of probing the activities of large and synthetically difficult structures.

Pyrethrum Flowers (Chrysanthemum cinerariaefolium)

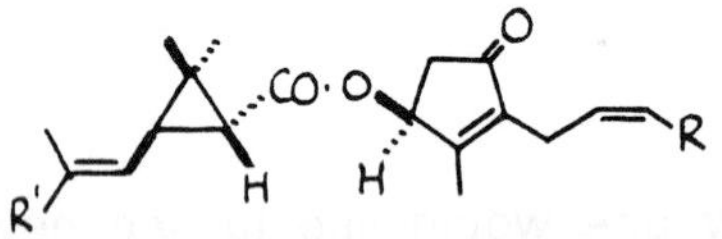

(1a) Pyrethrin I   R = CH=CH₂   R' = Me
(1b)   "   II   R = CH=CH₂   R' = CO₂Me
(1c) Cinerin I   R = Me   R' = Me
(1d)   "   II   R = Me   R' = CO₂Me
(1e) Jasmolin I   R = Et   R' = Me
(1f)   "   II   R = Et   R' = CO₂Me

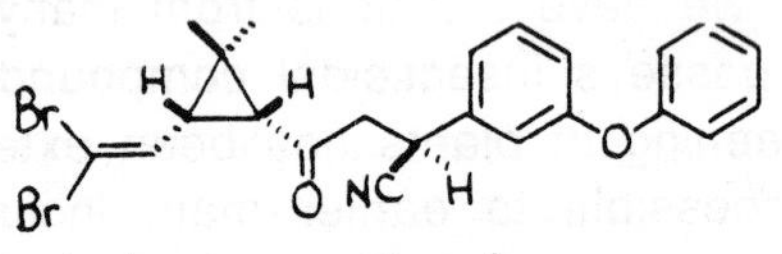

(1R)-cis-   (α-S)

Deltamethrin (2)

Natural insecticides of plant origin have been a major interest of our laboratory for many years. Our early studies concerned the stereochemistry and total synthesis of the natural pyrethrins (1a-1f), biosynthesised by Chrysanthemum cinerariaefolium.[1] The fact that they are esters readily factorised into two structures eases the problems of modelling

neo-Herculin

(3a)

Anacyclin

(3b)

Pipercide

(3c)

Rotenone (4)

and the development of the pyrethroid group of insecticides from these structures by Michael Elliott and his group at Rothamsted has been the outstanding advance of the insecticide area during the past 25 years culminating in compounds such as deltamethrin (2). Like some other natural insecticides the pyrethrins have insect repellent and antifeedant activity also. A second group of natural insecticide models with which we have had considerable involvement[2] has been the so-called isobutylamide group (e.g. 3a-3c) and the development of these compounds towards a commercial insecticide is the subject of Dr. Blade's contribution later in the Symposium. Rotenone (4) and the rotenoids have long been major subjects of study in Nottingham from stereochemical, synthetic, and biosynthetic angles.[3] In the shape of derris preparations, because of its status as a botanical insecticide, there seems to be increasing interest from those who do not approve of synthetic insecticides, but it is also an interesting model for synthetic design.

The tobacco hornworm larva feeds preferentially on plants of the Solanaceae family but it will prefer to starve to death rather than eat the leaves of Nicandra physaloides. Force-fed it dies, and the leaf extracts have been shown to be antifeedant towards the larvae of a range of insect species.

Nicandra physaloides

Nic-7 (5)

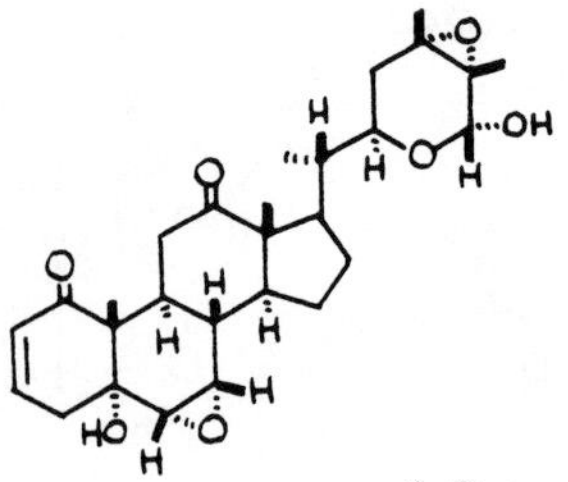

Nic-1 (6)

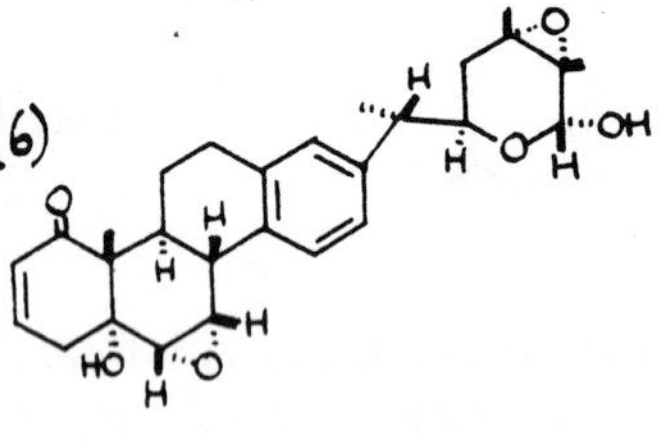

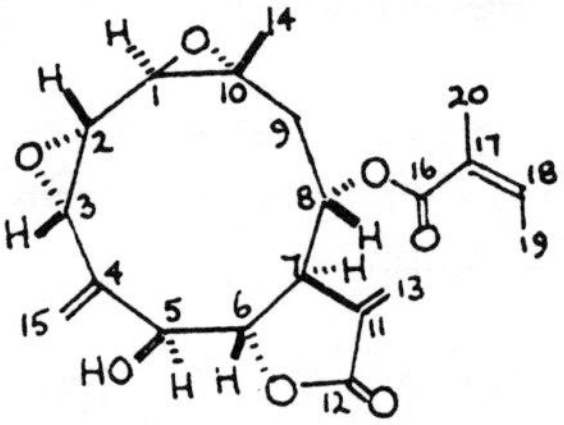

Cordifene (7)

Phorbol

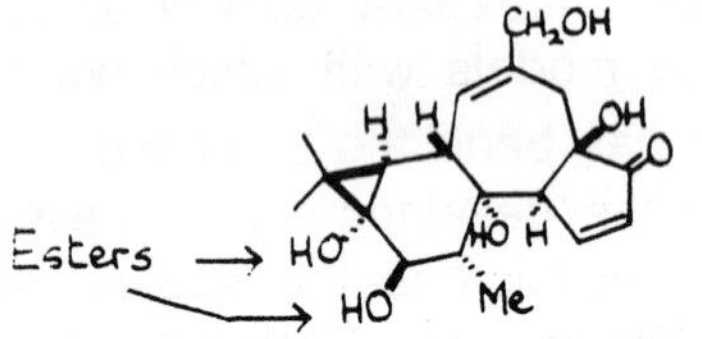

Esters →

(8)

Croton Oil Co-Carcinogens

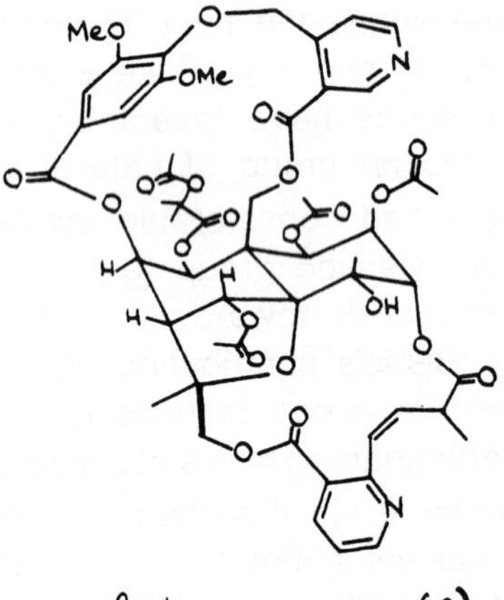

Cathedulin K-19 (9)

Our investigation[4] led to the isolation of three new steroids of the withanolide type (e.g. 'Nic-7' (5)) along with a family of four fascinating and novel steroids in which ring-D had been expanded and aromatised biosynthetically to incorporate the angular methyl group (e.g. 'Nic-1' (6)). Biological work with 'nicandrenone' which consists mainly of our 'Nic-1' indicates that the latter is the major antifeedant, though some activity doubtless resides in the other steroids present. We have also studied in detail the cordifene germacranolide group (e. g. cordifene (7)) from Erlangea cordifolia which are antifeedants to the army-worm and aphids.[5]

It was the reported insecticidal activity of phorbol esters (8) from Croton tiglium that was one factor that drew us to work in that area,[6] though the co-carcinogenicity of such compounds makes one cautious of their status as models. We were led into a long study of Euonymus alkaloids, mainly from the drug Khat(from Catha edulis) which contains many alkaloidal sesquiterpene esters such as catheduline K-19 (9),[7] partly through the insecticidal activity of members such as spindle (Euonymus europaeus) and the Chinese 'thunder-god vine' (Tripterygium willfordii): these contain similar, if rather simpler, compounds. Termite repellent triterpenes have also interested us.[8] Some of these models are complicated to explore by total synthesis and for the major part of this article I want to consider the mammea coumarins as natural insectides and potential models for further synthetic development.

The mammey tree, Mammea americana, is a member of the Guttiferae. It is a Caribbean tree indigenous to the West Indies, forming handsome evergreen specimens growing to 60 - 80ft . It has glossy elliptical dark green leaves and small white fragrant flowers, but it is not hardy and cannot withstand winter frosts. However it can be grown in Florida, and it has been introduced into Africa where a similar tree Mammea africana is indigenous. Our concern is with the fruit, the so-called 'mammey apple' or 'St. Domingo apricot', the flesh of which tastes like apricot and can be used either fresh or in preserves. However the flesh has long been suspected of being somewhat toxic, and on purchasing fruit in Guyana I was warned by my academic companion of eating the fruit and drinking rum at the same time. The flesh of the fruit contains usually one large seed (can be up to four) contained in a rough husk. The seed itself is about the size of a small hen's egg and is powdered and used as a traditional insecticide in the W. Indies

and Mexico against lice, fleas and sheep ticks. Such
preparations are also reported to be effective against
armyworms, mealworm, diamond-black moth larvae,
cockroaches, ants and mosquitoes.

Mammein

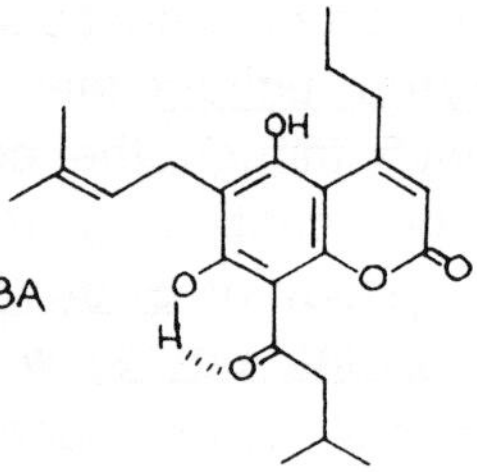

Mammea B/BA
(10)

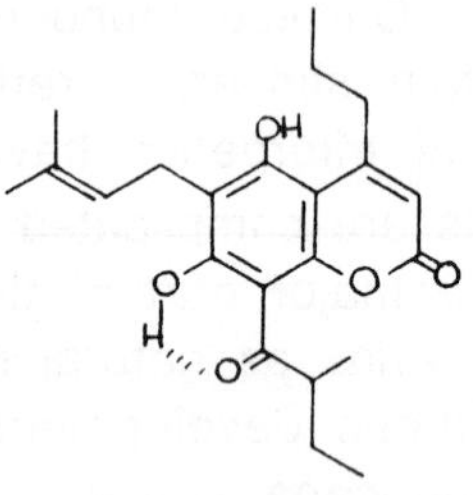

B/BB (11)

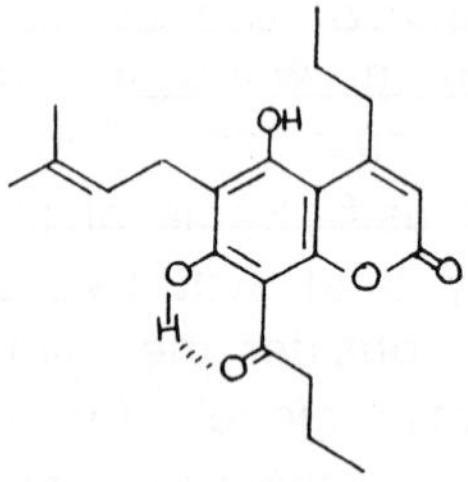

B/BC (12)

The first important work on the insecticidal principles of
M. americana was undertaken by Djerassi, Finnegan and their
colleagues in the late1950's and early 1960's.[9]   A crystalline
compound named 'mammein' (we now code it B/BA (10)) was
isolated and thought to be the insecticidal principle.   However
tests at Rothamsted against mustard beetle showed that it had
no significant activity and our examination showed that it was
contaminated with two difficultly separable companion
substances, mammea B/BB (11) and B/BC (12).   Crude extracts
of mammea seeds were confirmed as being active against
mustard beetles and so we began to search for the insecticidal
principle.

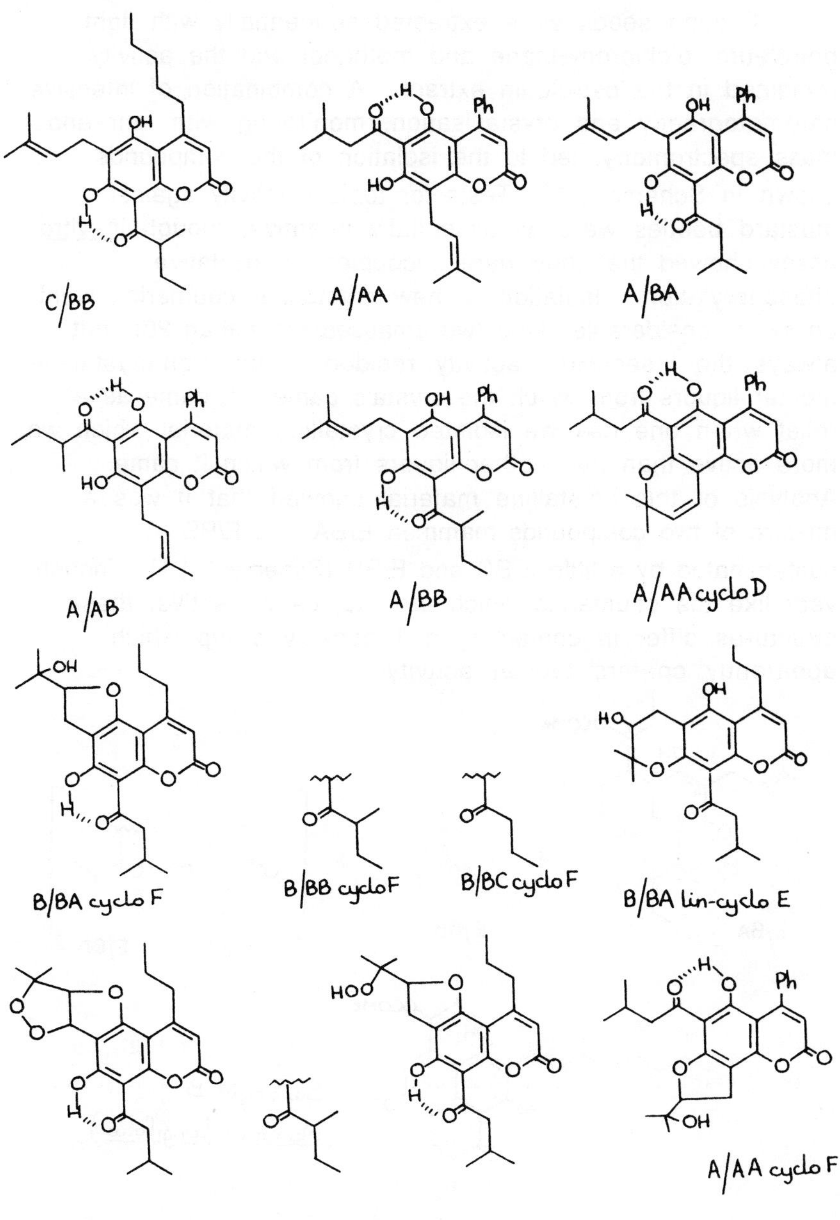

C/BB   A/AA   A/BA

A/AB   A/BB   A/AA cyclo D

B/BA cyclo F   B/BB cyclo F   B/BC cyclo F   B/BA lin-cyclo E

A/AA cyclo F

Scheme 1. Further Coumarins from Mammea americana.

   Ground seeds were extracted sequentially with light
petroleum, dichloromethane and methanol and the activity
remained in the petroleum extract.   A combination of intensive
chromatography and crystallisation, monitoring with nmr and
mass spectrometry, led to the isolation of the compounds
shown in Scheme 1.[10]   Tests for topical activity against
mustard beetles were all essentially negative, though in vitro
assay showed that they were uncouplers of oxidative
phosphorylation.   Isolation of new crystalline coumarins went
on for a considerable time (we amassed more than 20) but
always the insecticidal activity resided in the non-crystalline
mother liquors from which the crystals came.   It came as a
relief when one day we isolated crystalline material which was
more active than the mother liquors from which it came.
Analysis of this crystalline material showed that it was a
mixture of two compounds mammea E/BA and E/BB
contaminated by a little E/BC and E/BD (Scheme 2).[11]   Though
very like the coumarins which are   topically inactive, these
structures differ in containing a 1'-acetoxy group which
apparently confers topical activity.

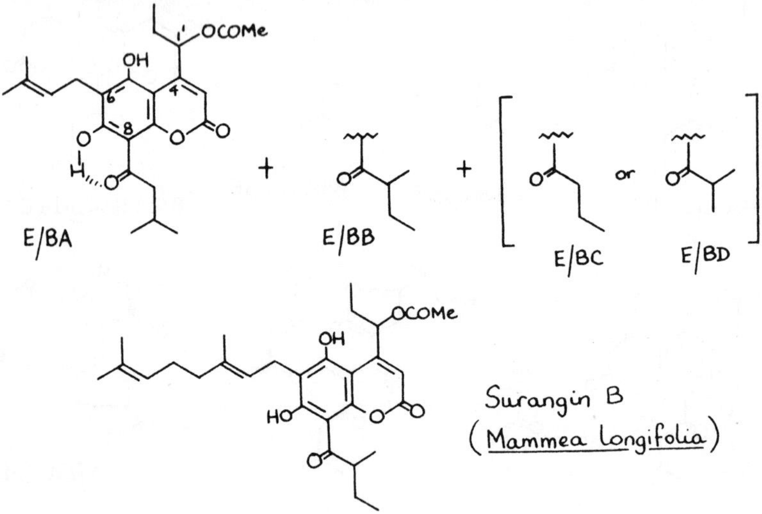

Scheme 2. Insecticidal 1'-Acetoxy Coumarins.

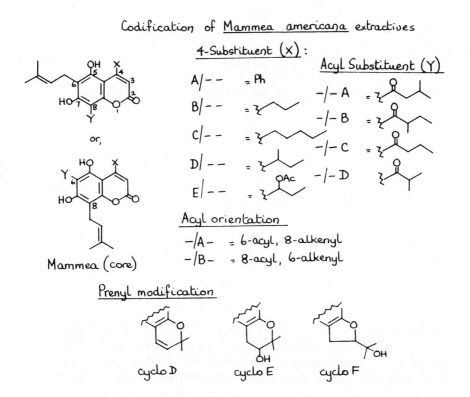

Scheme 3. Coding System for Mammea and Other Coumarins.

We were able to make a test of this deduction because Indian workers at about this time isolated a coumarin surangin B from Mammea longifola which also contained the 1'-acetoxy feature.[12] No insecticidal data had been reported so we tested a sample and were pleased to find that it was topically effective as an insecticide - it was in fact rather more active than our M. americana compounds.

In dealing with these mammea coumarins we resisted the temptation to proliferate trivial names for a large number of closely related compounds from one plant source and as indicated above we used a letter coding system. The key to this is given in Scheme 3.

It was clear at this stage that further progress now

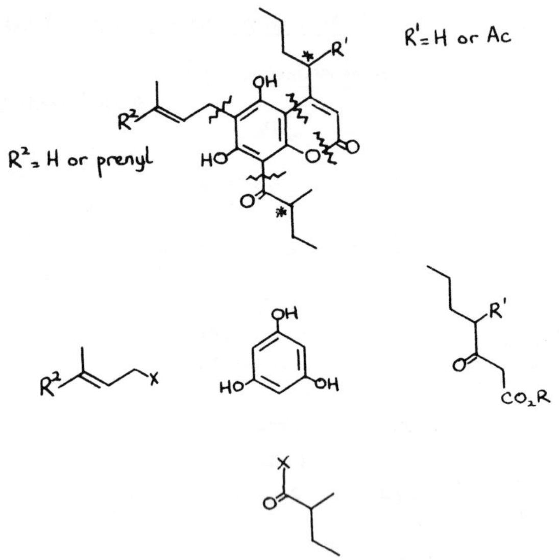

Scheme 4. Mammea Coumarins: Retro-Synthesis.

required a synthetic effort as the coumarins were available in only small amount and being so closely related were difficult to purify. Though fairly small molecules, they are quite densely functionalised and are sterically compressed: some molecules have two chiral centres. An obvious synthetic scheme is indicated (Scheme 4) and assuming that the coumarin ring is inserted in one step by the Pechmann reaction, there are six possible orders for insertion of the coumarin ring. Order is important since acylation deactivates the aromatic ring. Acid conditions can cyclise the 5-hydroxy with 1'-acetoxy substituents and cause deacylation: on the other hand, apart from obvious base hydrolysis, base conditions can also cause deacylation and induce an 8-acyl to 6-acyl rearrangement. 4-Substitution in a beta-ketoester can also cause difficulties in the Pechmann reaction. Prenylation also has its problems in a system of this kind. Unlike enolate anions where C-alkylation generally predominates, competition between O- and C-alkylation is more evenly balanced in phenolate anions and C-alkylation is burdened energetically since aromaticity is initially destroyed in the process.

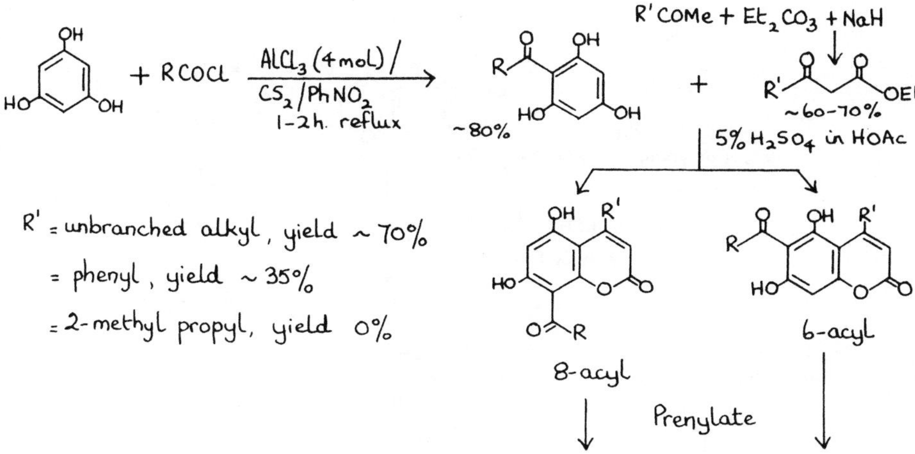

R' = unbranched alkyl, yield ~ 70%

= phenyl, yield ~ 35%

= 2-methyl propyl, yield 0%

Scheme 5. Acylation and Coumarin Ring Formation.

Earlier workers had acylated phloroglucinols in 20-40% yield, but yields seems to have been diminished through the use of inadequate amounts (1.5 mol) of aluminium chloride (Scheme 5). These acylated phloroglucinols underwent Pechmann condensation with gamma-substituted acetoacetates (Scheme 5) to give a mixture of 8-acyl and 6-acylcoumarins (~2:3), usually fairly readily separable by crystallisation from CHCl3/MeOH.[10] Yields with unbranched alkyl groups are very satisfactory, but in the case of R' = phenyl slow addition over 20 days was required to avoid excessive decomposition of ethyl benzoylacetate to acetophenone: branching proved very deleterious to the reaction. Our original orientations were based on the Gibbs' reaction (used circumspectly), verified later by an X-ray structure determination. For day to day control ultraviolet data with alkaline shifts provides a ready means of distinction between 6- and 8-acyl derivatives (Fig. 1). Prenylation was then studied as in Scheme 6.

Prenylation in such ambident phenolate anion systems is very solvent dependent. Potassium carbonate in acetone gives largely O-alkylation whereas in aqueous potassium hydroxide largely C-alkylation is observed, the negatively charged oxygen being screened by intense solvation leading to attack at carbon.

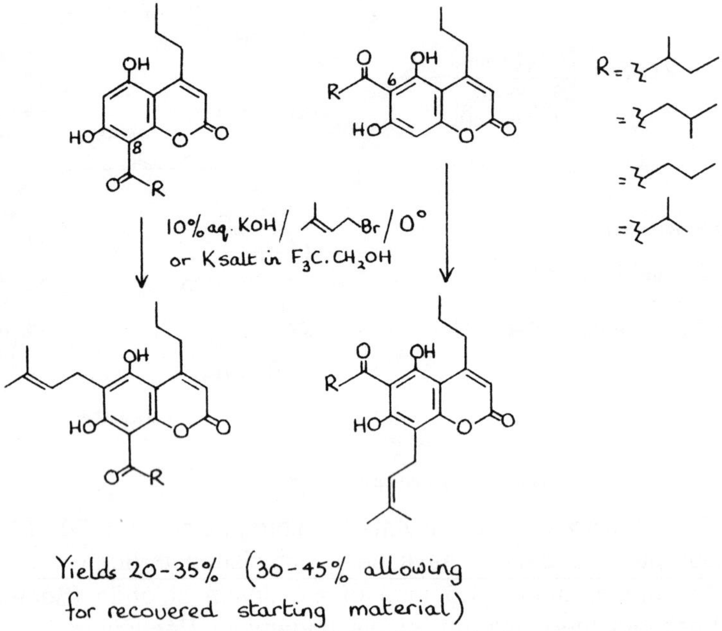

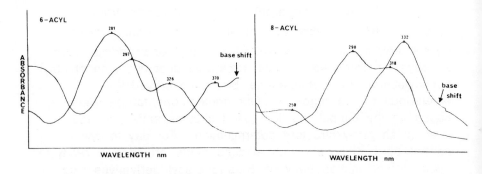

Yields 20-35% (30-45% allowing
for recovered starting material)

Scheme 6. Prenylation of Acylcoumarins.

Fig. 1. Typical Ultraviolet Data for 6- and 8-Acylcoumarins.

Using 10% potassium hydroxide at 0°C or the potassium salt of the phenol in the highly solvating trifluoroethanol, moderate yields were obtained in both the 8- and the 6-acyl series. All

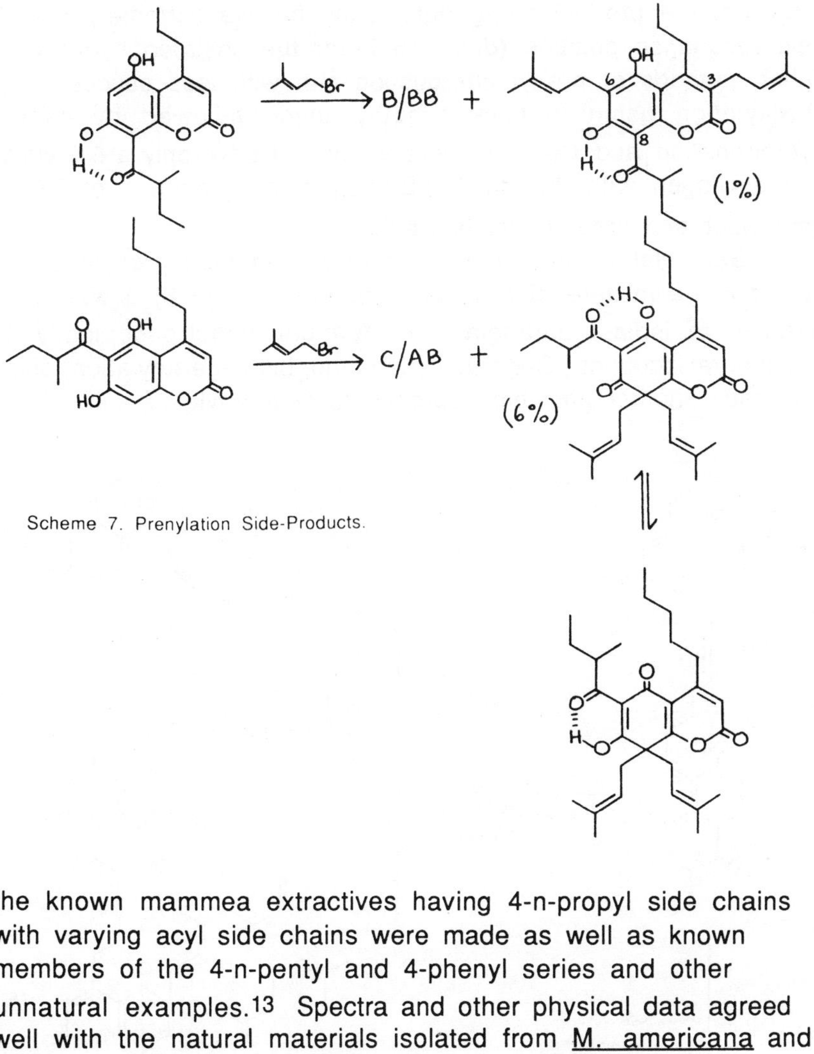

Scheme 7. Prenylation Side-Products.

the known mammea extractives having 4-n-propyl side chains
with varying acyl side chains were made as well as known
members of the 4-n-pentyl and 4-phenyl series and other
unnatural examples.[13] Spectra and other physical data agreed
well with the natural materials isolated from M. americana and
gave us very pure specimens free from closely similar
contaminents.

One or two side products obtained in the prenylation are
worthy of mention (Scheme 7). Minor competitive C-alkylation
can occur at the 3-position, a stabilised resonance position for
the phenolate anion, while in another example bis-prenylation

occurred, the product being noteworthy for the extremely low field resonance position (delta 18.5) for the chelated hydroxyl. Other methods of prenyl introduction met with less success. O-Alkylation using 3-chloro-3-methylbutyne followed by semi-hydrogenation and Claisen rearrangement[14] gave only a 6% yield and alkylation with 3-methylbut-2-enol in the presence of BF_3 gave poor and unreproducible yields.

Geranylation proceeded in poorer yield than prenylation, but from the mixture of products formed (Scheme 8) it was possible to isolate surangin A[12] identical spectroscopically with the natural product. O-Geranylation and bis-C-geranylation also occurred. Another possible approach to C-geranylated and

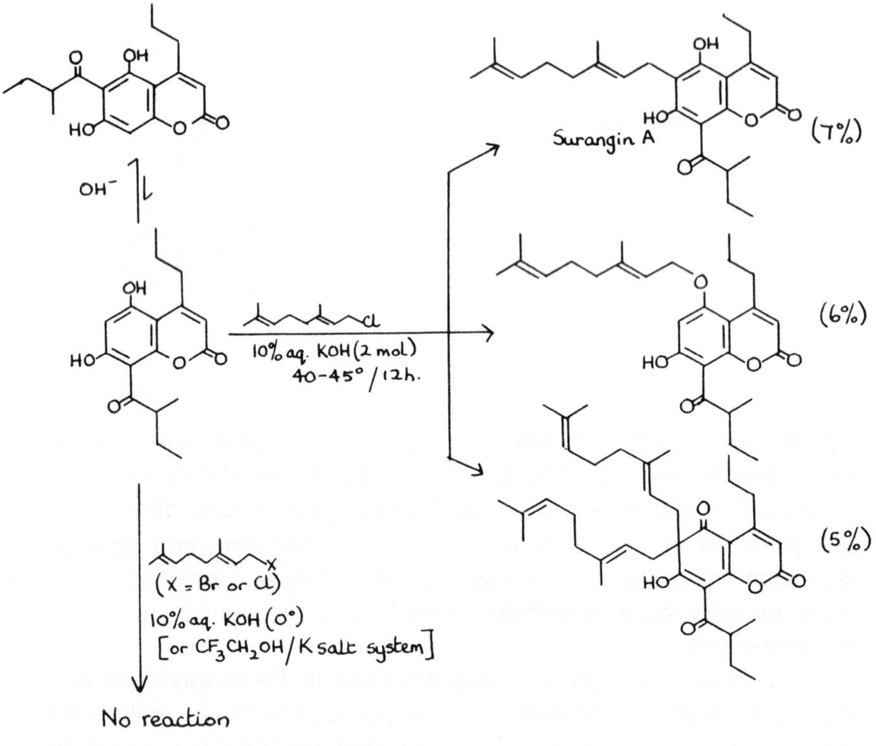

Scheme 8. Synthesis of Surangin A.

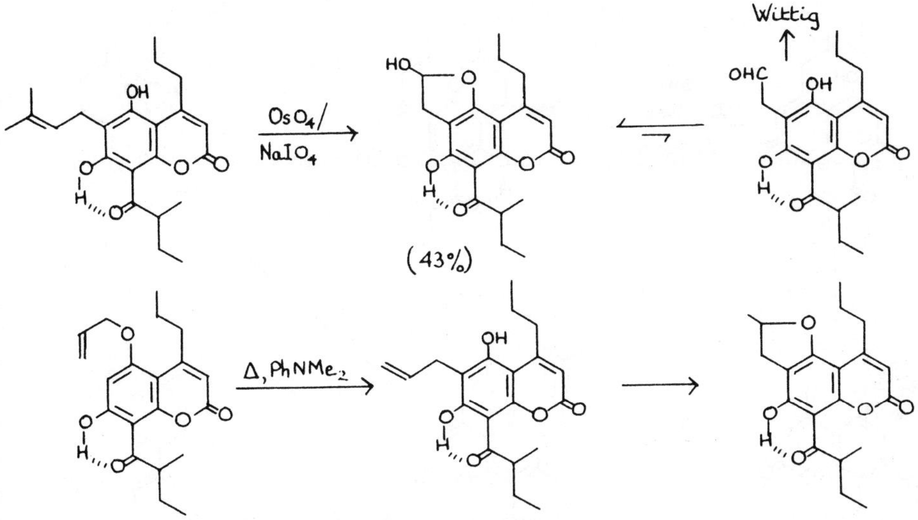

Scheme 9. Possible Wittig Approach to Geranyl Side-Chains.

similar products is via degradation of the prenyl to an aldehyde
followed by Wittig elaboration.   Constructing the aldehyde by
introducing a C-allyl group through Claisen rearrangement
however gave a cyclised product (Scheme 9) and this approach
was not followed up.   For the cyclo-D series we used two
routes. The Schmidt 3-chloro-3-methylbutyne method gave poor
yields but the use of 3-hydroxy-3-methylbutyraldehyde
acetal,[15] a general chromenylation method introduced by us
some years ago,[16] gave excellent results (Scheme10).
Cyclisations to give the cyclo-F and cyclo-E series were
carried out via the epoxide,(Scheme 11) the direction of
cyclisation being controlled by the addition or non-addition of
para-toluenesulphonic acid.[13]
      We have looked at a number of ways of making 4
-acetoxybutyroacetic ester required as a reagent for making
topically active coumarins by the Pechmann procedure.[17]
Rathke methodology (Scheme 12),[18] whilst successful, gave
only moderate yields by routes which did not scale up well.

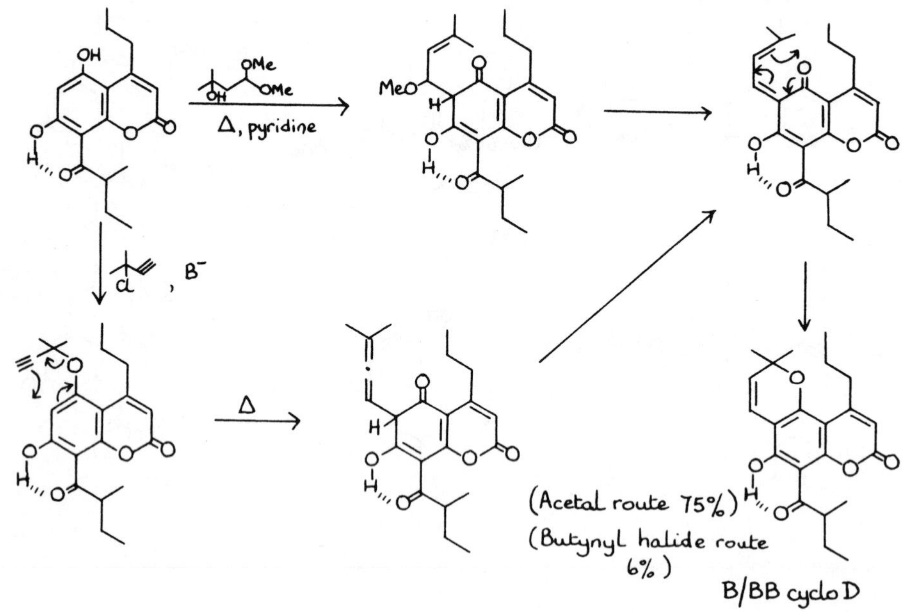

Scheme 10. Synthesis of Mammea Chromeno-Coumarins.

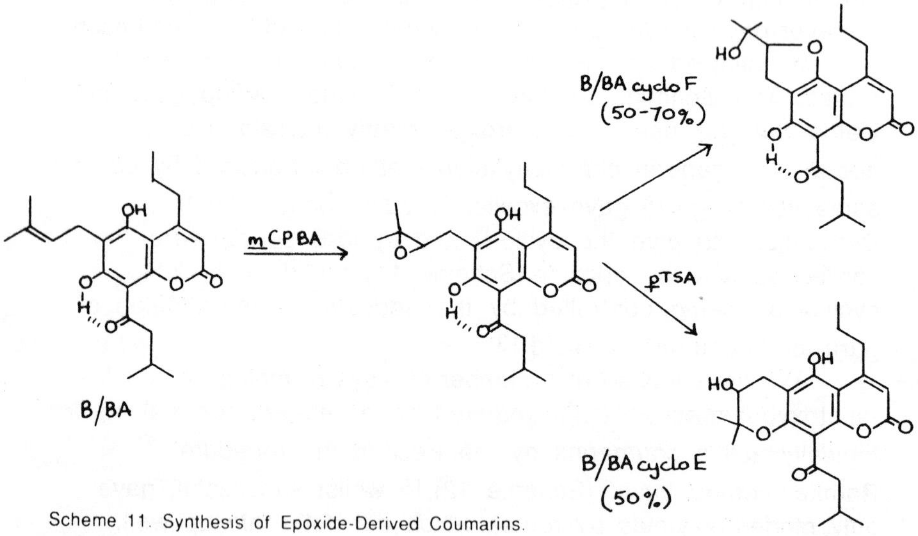

Scheme 11. Synthesis of Epoxide-Derived Coumarins.

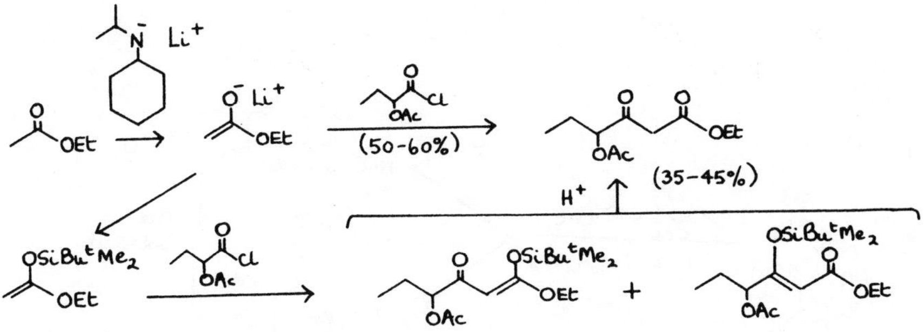

Scheme 12. Synthesis of 4-Acetoxybutyroacetic Ester by Rathke's Method.

Other methods were tried (Scheme 13) and the easiest and best route was the use of the magnesio-complex from malonic half ester[19] giving 88% yield overall. Unfortunately, despite several attempts, Pechmann reaction using the acetoxy substituted reagent was not successful (Scheme 14). Some

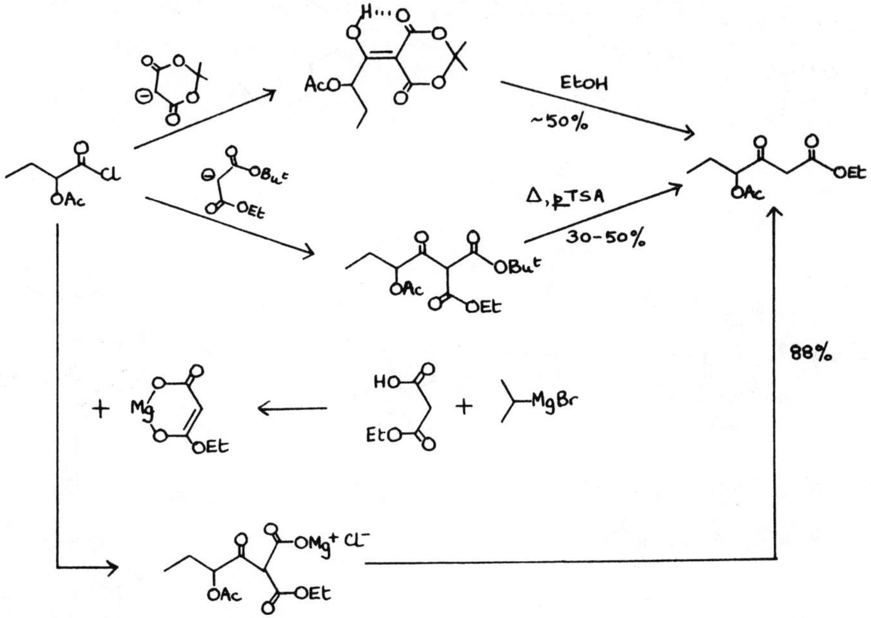

Scheme 13. Synthesis of 4-Acetoxybutyroacetic Ester by Other Methods.

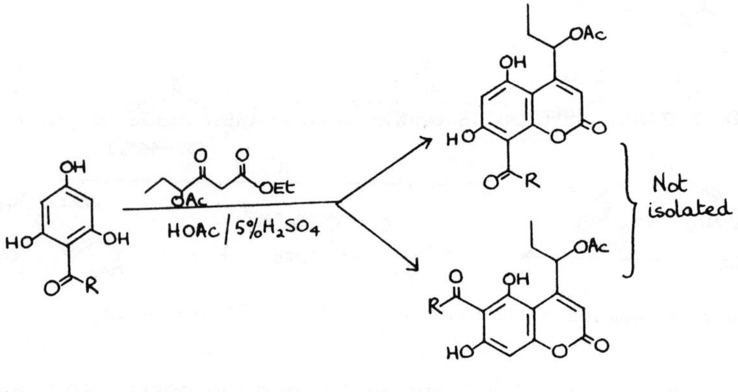

Scheme 14. Attempt to Acetoxycoumarinylate.

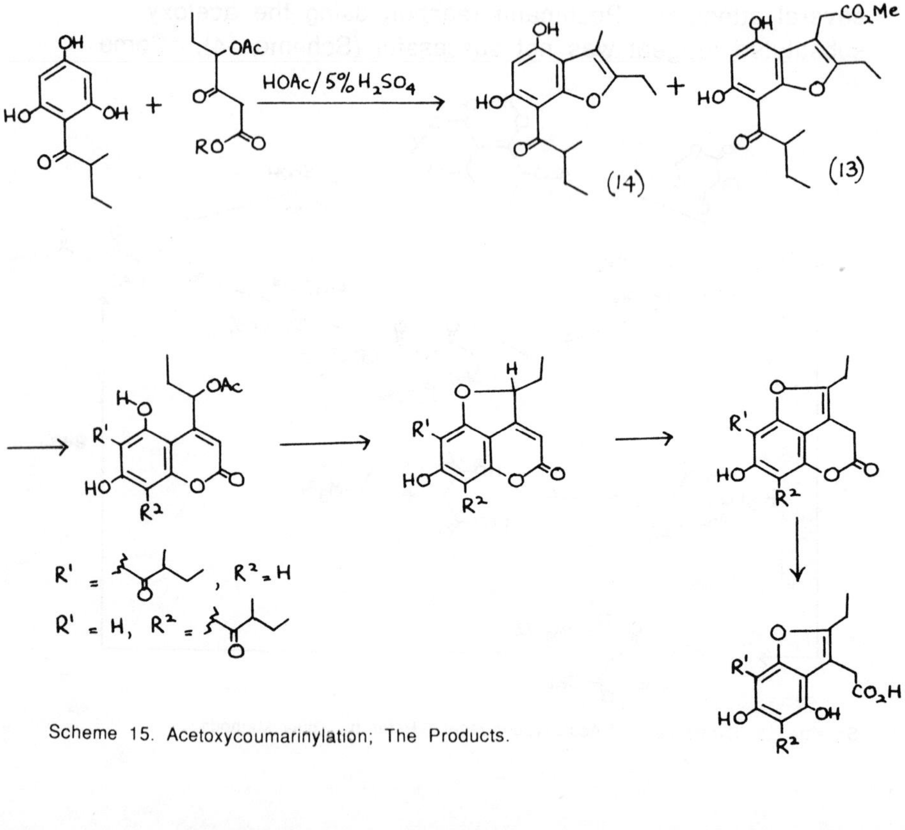

Scheme 15. Acetoxycoumarinylation; The Products.

products were isolated in low yield and give an indication of what is happening (Scheme 15). It would seem that the desired 1'-acetoxy compound is being formed but that cyclisation involving the 5-hydroxyl ensues followed by migration of the 3,4-double bond. Hydrolysis of the lactone then leads to (13), and (14) is formed by further decarboxylation. Other gamma-substituted beta-ketoesters were tried including the 4-oxo-compound, but without success. The 4-halo-series was looked at in some detail. Some of the desired coumarin (15) was obtained in poor yield along with considerable amounts of (16) and products deriving from dehalogenation and reduction when R' = Et and X = Br , though with R' = H and X = Cl a mixture of the 6- and 8- acyl compounds was isolated in 81% yield. Unfortunately attempts to displace the halogen led to cyclisation as in Scheme 16 .

Since the 5-hydroxyl was causing difficulties through cyclisation, it was acetylated along with the 7- in (17) and the 1'- bromine was introduced via N-bromosuccinimide . Again unwanted heterocyclisation caused difficulty (Scheme 17   ) so

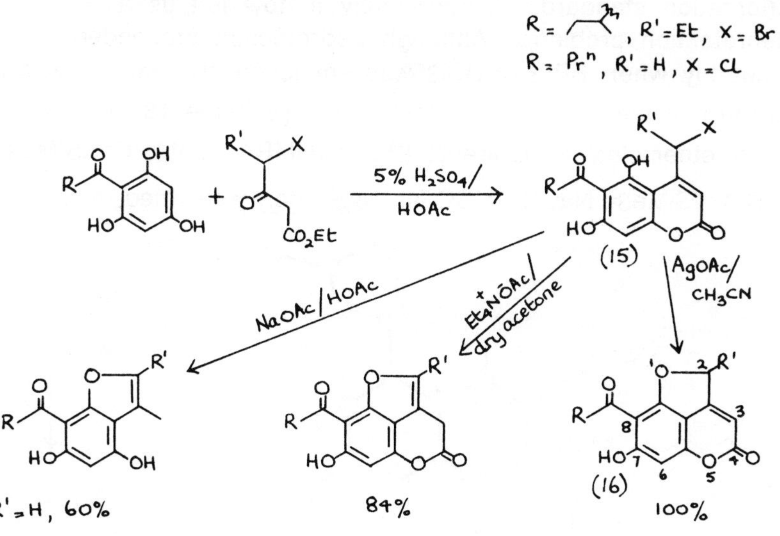

Scheme 16. Attempts at Halogen Displacement for 1'- Acetoxycoumarins.

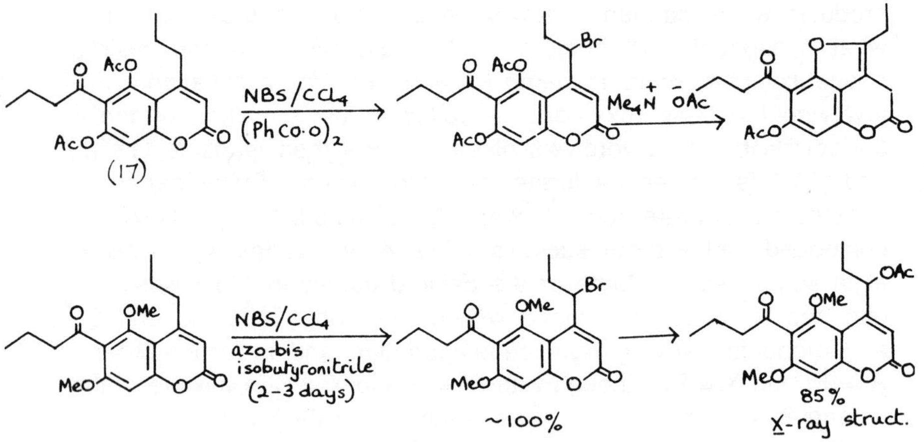

Scheme 17. Synthesis of Dimethylated 1'-Acetoxycoumarins.

the blocking group was changed to methyl. The 1'-acetoxy
compound could now be made in high yield and an X-ray
structure of the compound was carried out (Fig. 2)[20] as we
wished to have a coumarin of assured structure as an
orientation standard. Unfortunately it now left us a
deprotection problem. Although deprotection proceeded
smoothly when BBr3 or BCl3 was employed, the product was a
mixture of the two mono-methyl ethers (Scheme 18). Many
other ether cleaving reagents (But SLi/HMPA, MgI2, PhSSiMe3,
AlCl3, ISiMe3, NaCN/DMSO at 180° etc.) were tried, either

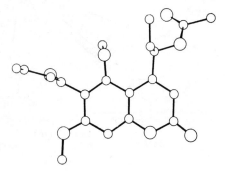

Fig 2. X-Ray Structure of the 6-Acyl 4-(1'-acetoxypropyl)coumarin
Dimethyl Ether.

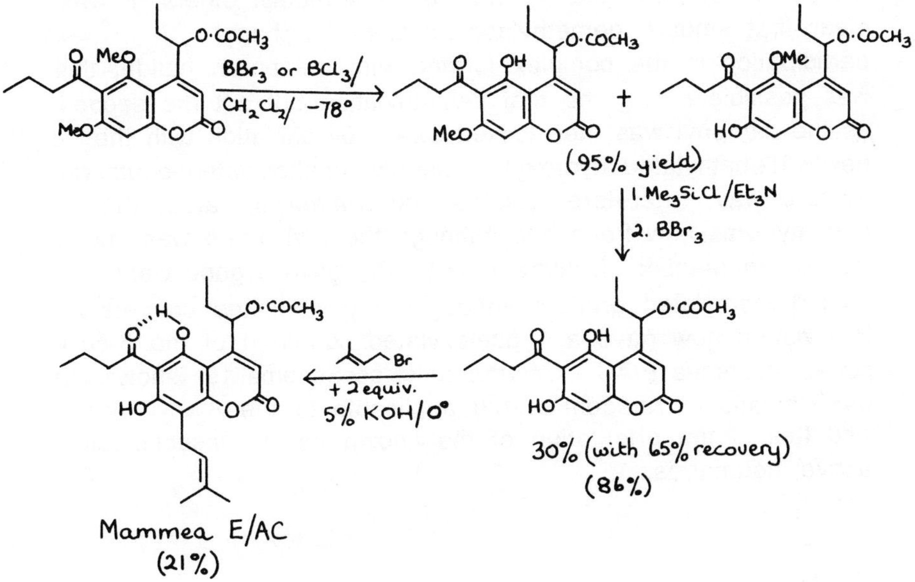

Scheme 18. Synthesis of Mammea E/AC.

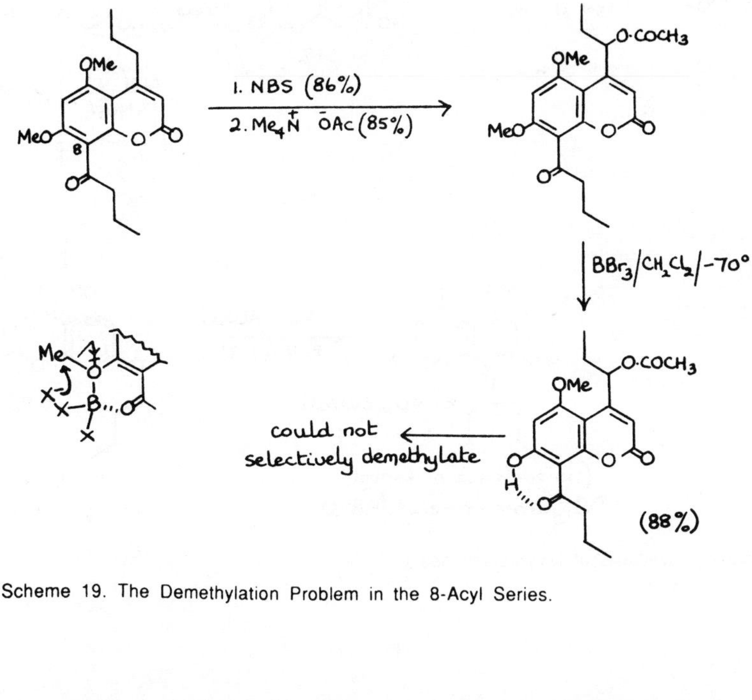

Scheme 19. The Demethylation Problem in the 8-Acyl Series.

without success or leading only to mono-methyl ethers. It was clear that smooth demethylation required carbonyl participation in the complex formed with the boron halide; this was possible during the first demethylation but not the second as the carbonyl was then locked away by chelation with the newly demethylated hydroxyl. The half methoxylated coumarin mixture was therefore isolated and trimethylsilylated; the carbonyl was now freed and although the silyl group was lost during the reaction, treatment with BBr$_3$ gave a good yield of fully demethylated product although only in limited conversion. Prenylation now gave a 1'-acetoxylated coumarin of the 6-acyl series, mammea E/AC. Unfortunately, as Scheme 19 shows the demethylation strategem is not applicable to the 8-acyl series and this is the orientation of the known natural insecticidally active coumarins.

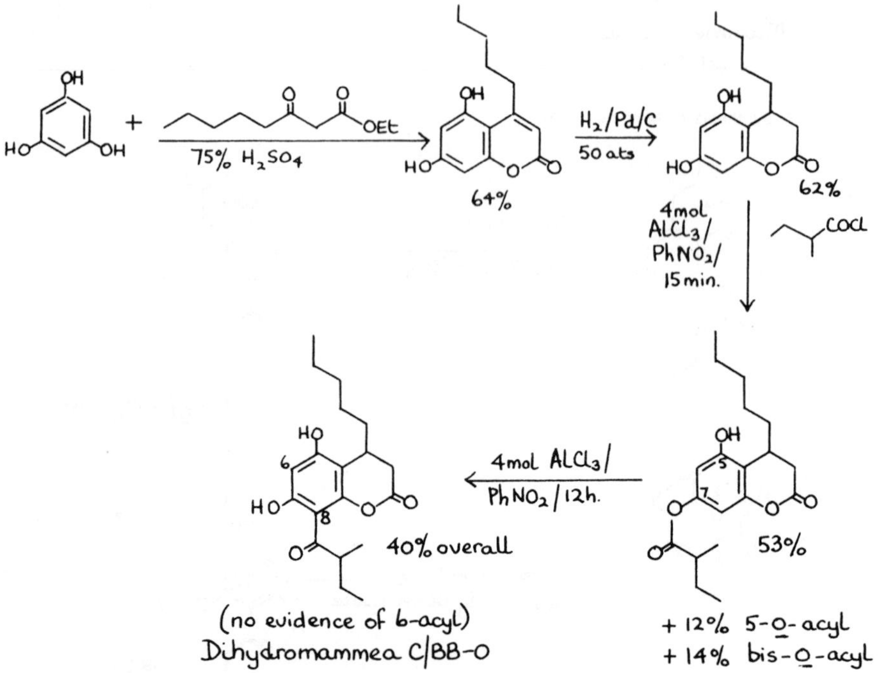

Scheme 20. Synthesis of Dihydromammea C/BB-0.

A different line of approach opened up during our synthesis[17] of the natural coumarin dihydromammea C/BB-O.[21] Here we inserted the coumarin ring first(Scheme 20), hydrogenated, and then acylated using aluminium chloride. The acylation apparently follows a Fries rearrangement pathway and it was highly specific for rearrangement to the 8-acyl compound. The same approach was used to make the natural 8-acyl coumarin mammea D/BB and again acylation was highly positionally specific for the 8-isomer (Scheme 21). This route was now adapted for the natural insecticidal coumarins.

Phloroglucinol itself reacted smoothly with the acetoxy beta-keto ester in trifluoroacetic acid to give the 1'-acetoxy-coumarin (18) which was acylated with the 2- and with 3-methylbutyryl chlorides. In each case 8-acyl products were formed with only small quantities of 6-. The products were prenylated to give the naturally occurring mammeas E/BB and E/BA in optically unresolved form (Scheme 22). A similar

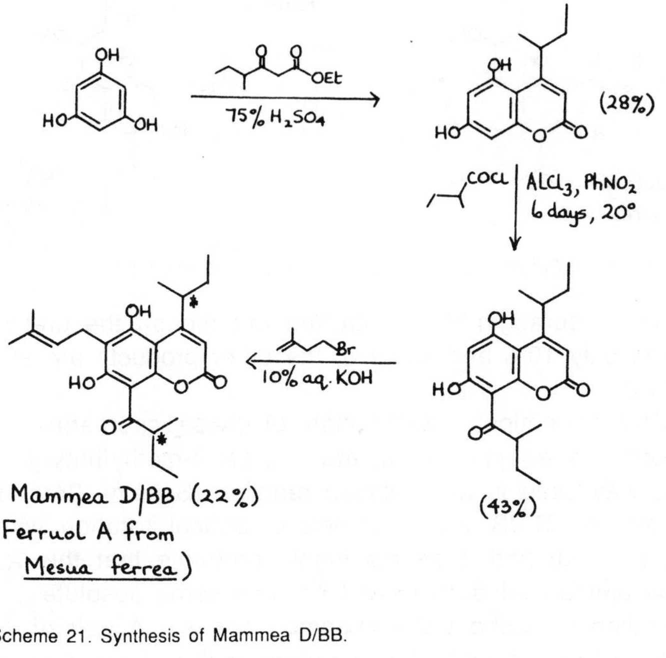

Mammea D/BB (22%)
(Ferruol A from
   Mesua ferrea)

(43%)

Scheme 21. Synthesis of Mammea D/BB.

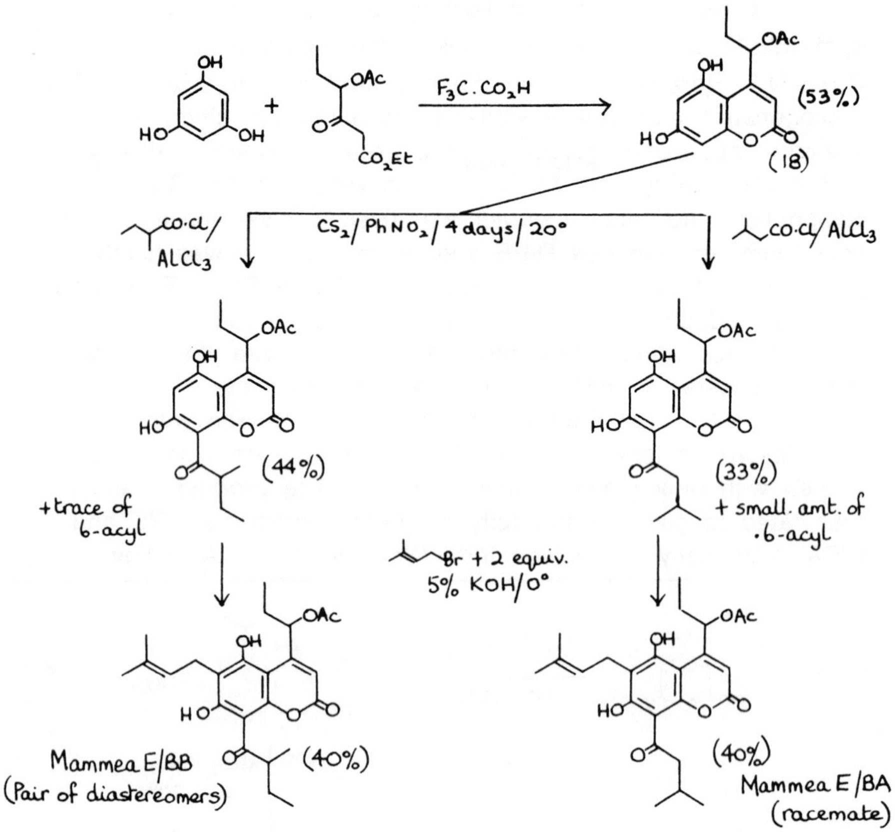

Scheme 22. Synthesis of Insecticidal Mammeas E/BB and E/BA.

synthesis of surangin B was carried out though the prenylation yield was only 10% and some of the other products are shown in Scheme 23 .

Stereochemical investigation of these coumarin compounds is as yet incomplete.   S-(+)-2-methylbutyryl chloride was used in an acylation reaction (Scheme 24   ) to give S-(-)-mammea B/BB which agreed in optical rotation with the natural material and it seems highly probable that this acyl group is amino-acid derived and has the same absolute configuration throughout the mammea series.   A pair of optically active surangin B diastereomers has been prepared

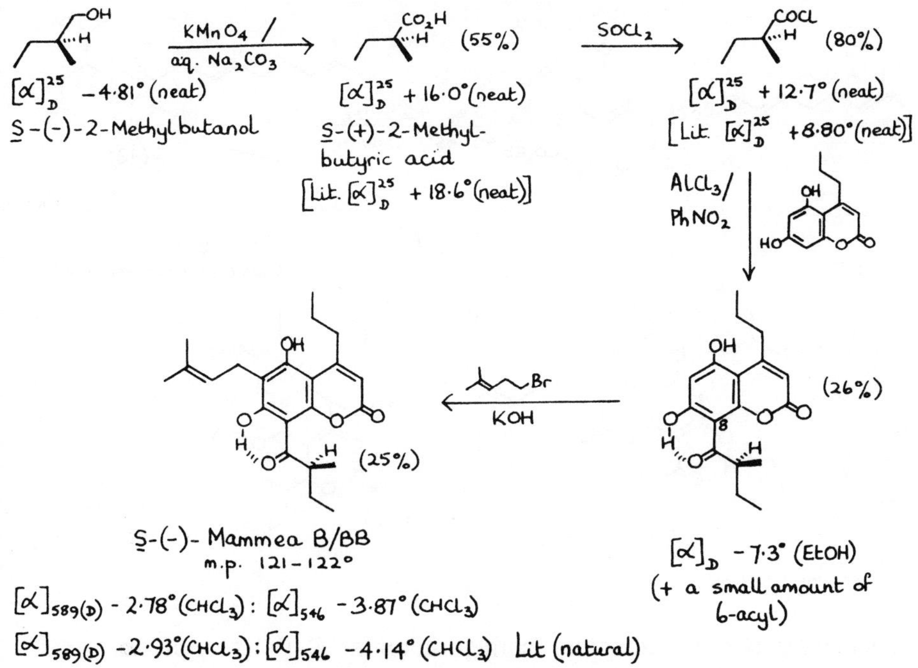

Scheme 24. Synthesis of S-(-)-Mammea B/BB.

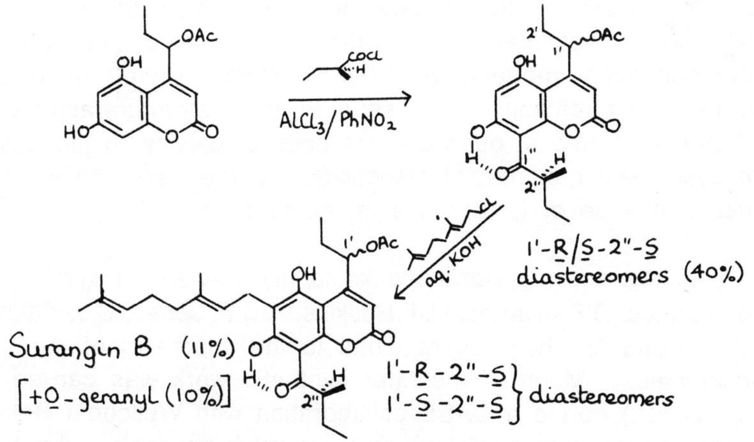

Scheme 25. Synthesis of Surangin B as Chiral Diastereomers.

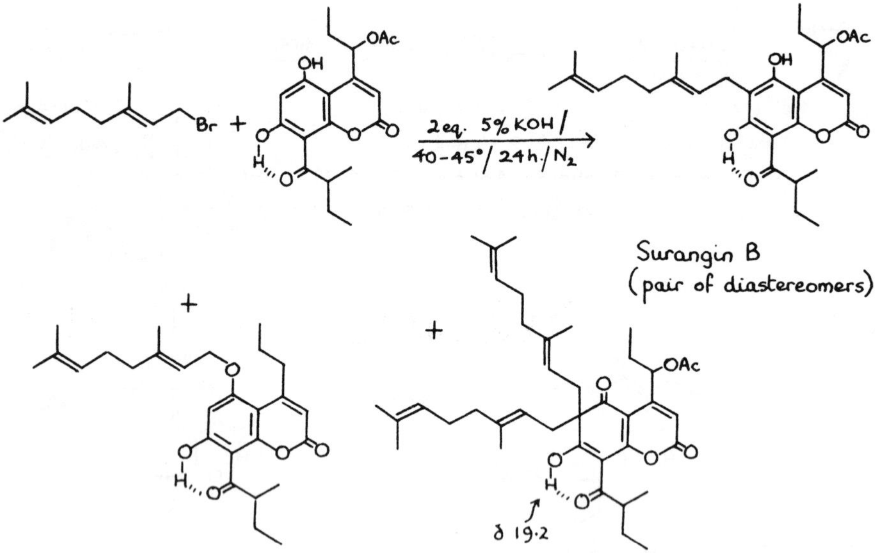

Scheme 23. Synthesis of Surangin B.

similarly but the absolute configuration at C-1' has not yet been determined (Scheme 25).

During work over many years it has not been our aim to try to develop commercially acceptable insecticides ourselves. Our aim has been to facilitate the development of new synthetic insecticides by providing natural structural and stereochemical models, and, by investigating their synthesis, to provide chemical tools useful in the structure/activity work of others. Some of our work has been of service to the design of successful commercial insecticides in the past and we hope that it may be of further use in the future.

I express my warm thanks to my co-workers and colleagues, D.E. Games, N.J. Haskins, R.C.F. Jones, C.J. Palmer and G.F. Reed for their contributions to the chemistry of the mammeins. Much of the later synthetic work was carried out (by C.J.P.) during a CASE collaboration with Wellcome Research Laboratories (Berkhamsted) and we much appreciate the keen interest of Dr. M.H. Black and Dr.J.B. Weston in its progress.

REFERENCES

1 For a summary see L. Crombie in 'Neurotox '88. Molecular Basis of Drug and Pesticide Action', Ed. G.G. Lunt. Elsevier Science Publishers, Amsterdam, New York and Oxford.

2 L. Crombie, M.A. Horsham and R.J. Blade, Tetrahedron Lett. 1987, 4879. L. Crombie, A.J.W. Hobbs, M.A. Horsham and R.J. Blade, ibid., 1987, 4875. L. Crombie and D. Fisher, ibid., 1985, 2477, 2481.L. Crombie and R. Denman, ibid., 1984, 4267. L. Crombie, A.H.A. Krasinski and M. Manzoor-i-Khuda, J. Chem. Soc., 1963, 4970. L. Crombie and M. Manzoor-i-Khuda, ibid., 1957, 2767. L. Crombie and J.L. Tayler, ibid. 1957, 2760. L. Crombie, ibid. 1955, 995, 1007.

3 For a biosynthetically oriented summary see L. Crombie, Natural Product Reports, 1984, 3.

4 M.J. Begley, L. Crombie, P.J. Ham and D.A. Whiting, J. Chem. Soc. Perkin Trans. 1, 1976, 296, 304

5 M.J. Begley, L. Crombie, W.M.L. Crombie, A.K. Gatuma and A. Maradufu, J. Chem. Soc., Perkin 1, 1981, 2702; 1984, 819.

6 L. Crombie, M.L. Games and D.J. Pointer, J. Chem Soc. (C), 1968,1347.

7L. Crombie. D. Toplis, D.A. Whiting, Z. Rozsa, J. Hohmann and K. Szendrei, J. Chem Soc. Perkin Trans.1, 1986, 531. R.L. Baxter, L. Crombie, D.J. Simmonds, D.A. Whiting, O.J. Braenden and K. Szendrei, ibid., 1979, 2965. R.L. Baxter, L. Crombie, D.J. Simmonds and D.A. Whiting, ibid., 1979, 2972. L. Crombie, W.M.L. Crombie, D.A. Whiting and K. Szendrei, ibid., 1979, 2976. R.L. Baxter, L. Crombie, W.M.L. Crombie, D.J. Simmonds, D.A. Whiting and K. Szendrei, ibid., 1979, 2982.

8 L. Crombie, P.J. Ham and D.A. Whiting, Chem. and Ind. (London), 1971,176.

9 C. Djerassi, E.J. Eisenbraun, R.A. Finnegan and R.B. Gilbert, J. Org. Chem., 1960, 25, 2164, 2169. R.A. Finnegan, M.P. Morris and C. Djerassi, ibid., 1961, 26, 1180. R.A. Finnegan, and W.H. Mueller, ibid., 1965, 30, 2342.

10 L. Crombie, D.E. Games and A. McCormick, J. Chem. Soc., (C), 1967, 2545, 2553. L. Crombie, D.E. Games, N.J. Haskins, G.F. Reed , R.A. Finnegan and K.E. Merkel, Tetrahedron Lett., 1970, 3975, 3979. L. Crombie, D.E. Games, N.J. Haskins and G.F. Reed, J. Chem. Soc., Perkin Trans. 1,1972, 2241,2248.

11 L. Crombie, D.E. Games N.J. Haskins and G.F. Reed, J. Chem. Soc.,Perkin Trans. 1, 1972, 2255.

12 B.S. Joshi, Y.N. Karnat, T.R. Govindachari and A.K. Ganguly, Tetrahedron, 1969, 25, 1453.

13 L. Crombie, R.C.F. Jones and C.J. Palmer, J. Chem. Soc., Perkin Trans. 1, 1987, 317.

14 R.D.H. Murray, M.M.Ballantyne and K.P. Mathai, Tetrahedron, 1971, 27, 1247.

15 D.E. Games and N.J. Haskins, J. Chem. Soc. Chem. Commun., 1971,1005.

16 W.M. Bandaranayake, L. Crombie and D.A. Whiting,J. Chem. Soc.(C), 1971, 811.

17 L. Crombie, R.C.F. Jones and C.J. Palmer, J. Chem. Soc.,Perkin Trans. 1, 1987, 333.

18 M.W. Rathke and J. Deitch, Tetrahedron Lett., 1971,

2953. M.W. Rathke and D.F. Sullivan, *ibid.*, 1973, 1297.

[19] P. Pollet and S. Gelin, *Tetrahedron*, 1978, **34**, 1453.

[20] The X-Ray structure was determined by Dr. M.J. Begley in our laboratory.

[21] E.G. Crichton and P.G. Waterman, *Phytochemistry*, 1978, **17**, 1783.

[22] L. Crombie, R.C.F. Jones and C.J. Palmer, *J. Chem. Soc., Perkin Trans. 1*, 1987, 345.

Novel Avermectin Insecticides and Miticides

M. H. Fisher

MERCK SHARP AND DOHME RESEARCH LABORATORIES, P.O. BOX 2000, RAHWAY, NEW JERSEY 07065, USA

In 1976 scientists at Merck & Co. Inc. discovered a complex of eight closely related natural products, subsequently named avermectins, in a culture of <u>Streptomyces avermitilis</u> MA-4680 (NRRL8165) originating from an isolate by the Kitasato Institute from a soil sample collected at Kawana, Ito City, Shizuoka Prefecture, Japan. Their structures are shown in Fig. 1.[1] They are among the most potent anthelmintic, insecticidal and acaricidal compounds known.

Fig. 1

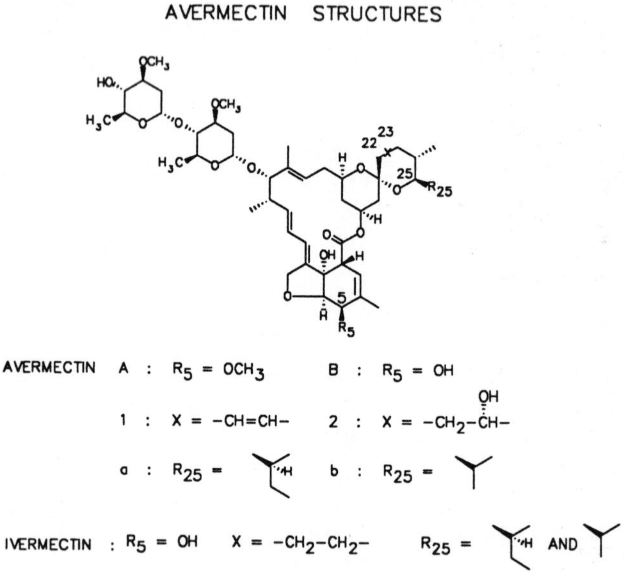

AVERMECTIN STRUCTURES

AVERMECTIN A : R_5 = OCH_3 B : R_5 = OH

1 : X = $-CH=CH-$ 2 : X = $-CH_2-CH-$
                                              OH

a : R_{25} =   b : R_{25} =

IVERMECTIN : R_5 = OH   X = $-CH_2-CH_2-$ R_{25} =   AND

The avermectins are closely related to another group of pesticidal natural products, the milbemycins, the first examples described by Japanese workers, but later found to be more abundant in nature than the avermectins.[2-5] Both the avermectins and milbemycins are sixteen-membered lactones, with a spiroketal system containing two six-membered rings. The principal difference is that the avermectins have an α-L-oleandrosyl-α-L-oleandrosyl disaccharide attached at the 13-position whereas the milbemycins have no 13-substitutent. Milbemycin structures are shown in Fig. 2.

Fig. 2

MILBEMYCIN STRUCTURES

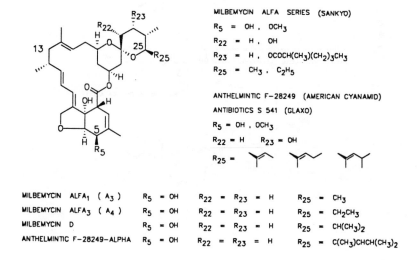

MILBEMYCIN ALFA SERIES (SANKYO)

R_5 = OH , OCH_3

R_{22} = H , OH

R_{23} = H , $OCOCH(CH_3)(CH_2)_3CH_3$

R_{25} = CH_3 , C_2H_5

ANTHELMINTIC F-28249 (AMERICAN CYANAMID)

ANTIBIOTICS S 541 (GLAXO)

R_5 = OH , OCH_3

R_{22} = H R_{23} = OH

MILBEMYCIN ALFA$_1$	(A$_3$)	R_5 = OH	R_{22} =	R_{23} = H	R_{25} = CH_3
MILBEMYCIN ALFA$_3$	(A$_4$)	R_5 = OH	R_{22} =	R_{23} = H	R_{25} = CH_2CH_3
MILBEMYCIN D		R_5 = OH	R_{22} =	R_{23} = H	R_{25} = $CH(CH_3)_2$
ANTHELMINTIC F-28249-ALPHA		R_5 = OH	R_{22} =	R_{23} = H	R_{25} = $C(CH_3)CHCH(CH_3)_2$

Two avermectins have been commercialized to date. Selective reduction of the 22,23-olefin of avermectin B1 yields the 22,23-dihydro derivative assigned the non-proprietary name ivermectin Fig. 3. Although this structure, for the sake of simplicity, depicts the 25-secbutyl derivative it should be noted that both commercial products contain up to 20% of the 25-isopropyl analog.

Fig. 3

SYNTHESIS OF IVERMECTIN

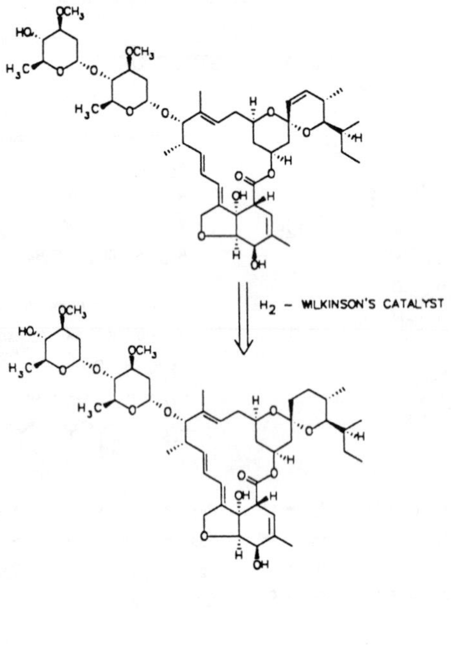

H₂ — WILKINSON'S CATALYST

A summary of the biological properties of ivermectin is shown in Fig. 4.

Fig. 4

IVERMECTIN

USED IN CATTLE AT 0.2 MG/KG
                SHEEP   0.2
                SWINE   0.3
                HORSES   0.2
                DOGS   0.006
                MAN   0.05 to 0.2

EFFECTIVE AGAINST PARASITIC NEMATODES
                             GRUBS
                             LICE
                             MITES
                             TICKS
                             BOTS

NOT ACTIVE AGAINST TAPEWORMS
                   FLATWORMS
                   BACTERIA
                   FUNGI

Avermectin B$_1$ is the most effective of the avermectin family of natural products against agriculturally important insects and mites. It has been commercialized for agricultural use under the non-proprietary name abamectin. A summary of its biological activity is shown in Fig. 5.

Fig. 5

ACTIVITY OF AVERMECTIN B₁ AGAINST MITES AND INSECTS

Mite Species (Contact effect against adult mites)	LC₉₀ (ppm)
Phyllocoptruta oleivora (citrus rust mite)	0.02
Tetranychus urticae (two-spotted spider mite)	0.03
Tetranychus turkestani (strawberry mite)	0.08
Panonychus ulmi (European red mite)	0.04
Panonychus citri (citrus red mite)	0.24
Polyphagotarsonemus latus (broad mite)	0.03
Insect Species (Foliar Residue Bioassay)	**LC₉₀ (ppm)**
Leptinotarsa decemlineata (Colorado potato beetle)	0.03
Manduca secta (tomato hornworm)	0.02
Epilachna varivestes (Mexican bean beetle)	0.20
Acyrthosiphon pisum (pea aphid)	0.40
Trichoplusia ni (cabbage looper)	1.0
Heliothis zea (corn earworm)	1.5
Spodoptera eridania (southern armyworm)	6.0

R. A. Dybas, A. St. J. Green (1984) Avermectins: Their Chemistry and Pesticidal Activity. Proceedings, 1984 British Crop Protection Conference-Pests and Diseases, Brighton, England, 31, 947-954.

Avermectin B₁, in thin films, is rapidly degraded on exposure to air and to ultraviolet light. In fact its utility against certain crops is limited by this rapid degradation. Fig. 6 shows the degradation of avermectin B₁ as a thin film on a glass petri dish cover held in the dark or exposed to a Kratos model LH 153 Solar Simulator. In the dark the degradative, processes are probably oxidative.

Fig. 6

PHOTODECOMPOSITION OF AVERMECTIN B_1

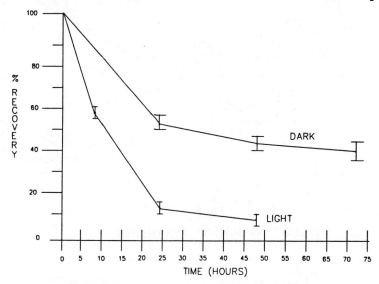

J.G. MacConnell, R.J. Demchak, F.A. Preiser, and R.A. Dybas, *J. Ag. Food Chem.* in press.

It has been shown for example, in our laboratories, that the 8a position is readily converted into a hydroperoxide. This reactivity is not unreasonable since the 8a position it both allylic and adjacent to an oxygen. When avermectin B_1 is dissolved in methanol or cyclohexane in a quartz tube and exposed to 300 nM ultraviolet light an equilibrium to the 8,9-$\underline{Z}$ and 10,11-$\underline{Z}$ isomers is achieved in 30 - 60 minutes Fig. 7 and complete loss of 254 nM absorption occurs in less than 24 hours. Mass spectral analysis of the products indicated up to four additional oxygen atoms.Since the early events in photodecomposition are related to ultraviolet absorption at the diene portion of the molecule it was decided to undertake chemistry at that site to hopefully improve photostability.

Since the early events in photodecomposition are related to ultraviolet absorption at the diene portion of the molecule it was decided to undertake chemistry at that site to hopefully improve photostability.

Hydrogenation of avermectin B_1 with hydrogen and a palladium catalyst gave 10,11,22,23-dihydro avermectin B_1 as shown in Fig. 8. Direct hydrogenation was studied with many catalysts and in no case could reduction of the diene be accomplished without prior or concomitant reduction at the 22,23-olefin.

Fig. 7

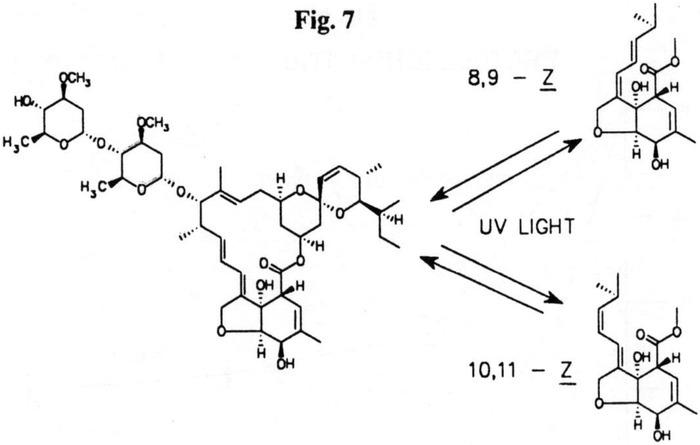

8,9 – Z

UV LIGHT

10,11 – Z

Fig. 8

HYDROGENATION OF AVERMECTIN B1a

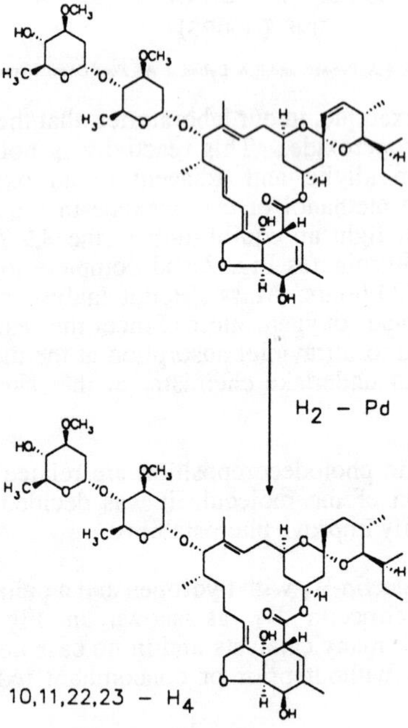

H_2 – Pd

10,11,22,23 – H_4

Selective reduction at the 10,11 position of the diene was accomplished by an indirect method shown in Fig. 9

Fig. 9

ADDITION OF N-BROMOACETAMIDE TO THE AVERMECTIN B1a - DIENE

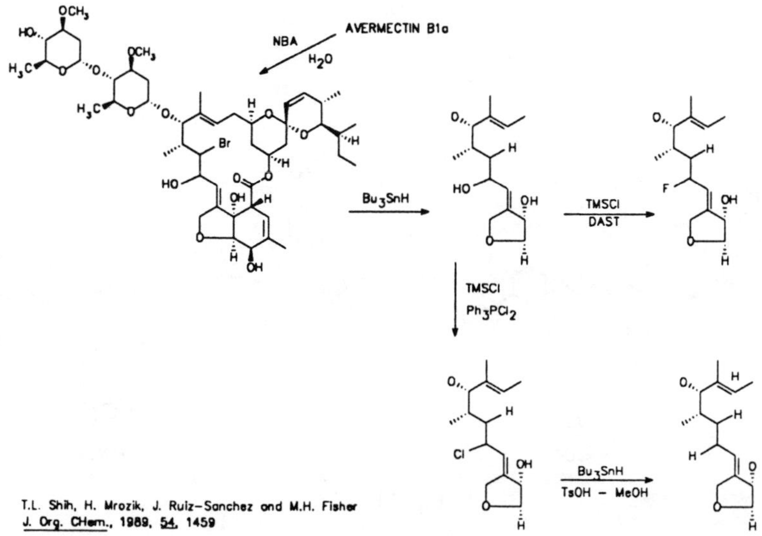
AVERMECTIN B1a
NBA
H_2O
Bu_3SnH
TMSCl
DAST
TMSCl
Ph_3PCl_2
Bu_3SnH
TsOH - MeOH
T.L. Shih, H. Mrozik, J. Ruiz-Sanchez and M.H. Fisher
J. Org. Chem., 1989, 54, 1459

Reaction of avermectin B_1 with N-bromoacetamide afforded a 10,11-bromohydrin which was reduced with tributyltin hydride to a 10-hydroxy derivative. This alcohol was protected at the 5-position and converted into the 10-chloro and 10-fluoro analogs. Reduction of the 10-chloro derivative with tributyltin hydride and deprotection gave the desired 10,11-dihydro avermectin B_1.

Epoxidation of avermectin B_1 with MCPBA gave predominantly the 8,9-oxide as shown in Fig. 10, together with a small amount fo the 3,4-oxide. Presumably both reactions are assisted by the 7-α-hydroxy group.

Reaction of 5-0-TBDMS ivermectin with a zinc/copper couple and methylene iodide in ether gave a mixture of three compounds shown in Fig. 11. Two monocyclopropanes were isolated in which reaction occurred at the α-face. Interestingly when the reaction was carried out with the unblocked 5-alcohol, the same products were formed. Presumably the stereochemistry is entirely controlled by the 7-hydroxy group as was expoxidation.

Fig. 10
EPOXIDATION OF AVERMECTIN B1a

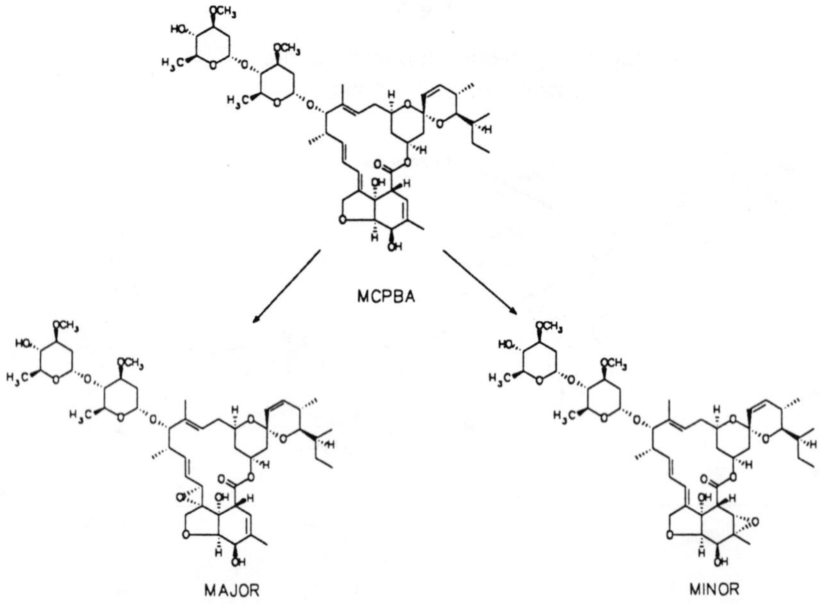

MCPBA

MAJOR   MINOR

Fig. 11

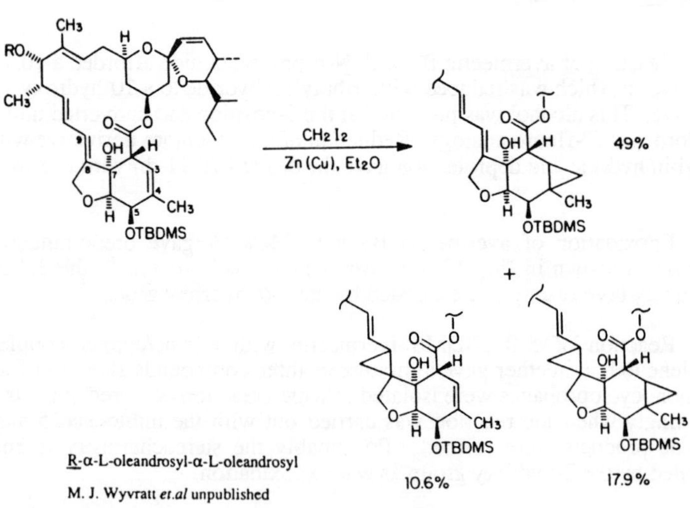

$CH_2 I_2$
$Zn (Cu), Et_2O$

49%

+

R-α-L-oleandrosyl-α-L-oleandrosyl

M. J. Wyvratt *et.al* unpublished

10.6%   17.9%

The miticidal activity of these derivatives is shown in Fig. 12.

Fig. 12

CONTACT ACTIVITY OF AVERMECTIN DERIVATIVES
AGAINST
TWO-SPOTTED SPIDER MITE **ADULT FEMALES**

Compound	Percent mortality at 96 hours 0.05 ppm
Avermectin (AVM B$_1$)	100
AVM B$_1$ 8,9-oxide	100
AVM B$_1$ 8,9-cyclopropane	15
AVM B$_1$ 3,4-cyclopropane	20
10,11-dihydro AVM B$_1$	100
22,23-dihydro AVM B$_1$ (Ivermectin)	92
10,11,22,23-tetrahydro AVM B$_1$	100
3,4,10,11,22,23-hexahydro AVM B$_1$	11
3,4,8,9,10,11,22,23-octahydro AVM B$_1$	18
10-fluoro-10,11-dihydro AVM B$_1$	100
10-hydroxy-10,11-dihydro AVM B$_1$	72
Milbemycin (25-sec-butyl) 8,9-oxide	20

Several of the reduced diene derivatives were found to be highly active. Interestingly avermectin B_1-8,9-oxide was highly active whereas the analogous cyclopropane was virtually inactive. Photodecomposition studies in a photoreactor Fig. 13 and as thin films on petri dishes Fig. 14 showed both the 8,9-oxide and the 8,9-cyclopropane to be considerably more stable than the parent.

Fig. 13

PHOTO DECOMPOSITION STUDIES

300 NM UV LIGHT

QUARTZ TUBES

MeOH or CYCLOHEXANE SOLUTIONS

	ISOMERIZATION EQUILIBRIUM
	30 - 60 MINUTES
AVM - B_1	
	COMPLETE LOSS OF 254 NM UV
	ABSORPTION IN LESS THAN 24 HOURS

8,9-METHYLENE-B_1 50% LEFT AFTER 24 HOURS

B_1-8,9-OXIDE   30 TO 50% LEFT AFTER 24 HOURS

The most active derivatives were also for foliar residual activity against two-spotted spider mites. Three derivatives, avermectin B_1-8,9-oxide, 10-11-dihydroavermectin B_1 and 10-fluoro-10,11-dihydroavermectin B_1 showed much improved residual activity when compared to avermectin B_1, Fig. 15.

Avermectin B_1-8,9-oxide has been selected for further study partially because it appeared to be the most effective compound but also because of its ease of synthesis.

Fig. 14
COMPARATIVE PHOTODECOMPOSITION OF AVERMECTIN B1a & IT'S 8,9-OXIDE

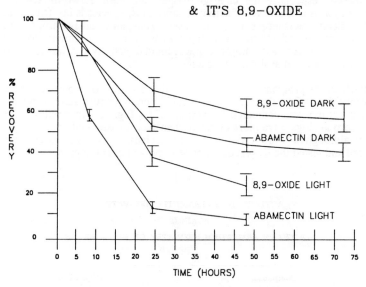

J.G. MacConnell, R.J. Demchak, F.A. Preiser, and R.A. Dybas, *J. Ag. Food Chem.* in press.

Fig. 15
FOLIAR RESIDUAL ACTIVITY OF
AVERMECTIN DERIVATIVES
AGAINST
TWO-SPOTTED SPIDER MITES ADULT FEMALES

	Percent mortality at 0.1 ppm	
Compound	**0 DAT[a]**	**15 DAT[a]**
Avermectin B₁(AVM B₁)	96.2	16.9
AVM B₁ 8,9-oxide	99.5	70.7
10,11-dihydro AVM B₁	98.0	67.0
10,11-22,23-tetrahydro AVM B₁	95.1	< 5
10-fluoro-10,11-dihydro AVM B₁	92.3	60.2

[a] 0 DAT and 15 DAT = 0 and 15 days after treatment spider mites placed onto foliage; mortality counts made 96 hours after infestation

Inspection of biodata shown in Fig. 5 indicates that whereas avermectin B_1 is extremely effective against a variety of mites, it is much less effective against insects, especially the cabbage looper, the corn earworm and the southern armyworm. The level of activity against these species is insufficient to justify commercial development for these uses. Thus, an extensive program of synthetic chemistry and biological testing was initiated in an attempt to find avermectin derivatives with improved insecticidal activity. The southern armyworm was selected as the target species.

Early in the program it was discovered that a variety of monosaccharide and aglycone derivatives showed a sixteen-fold improvement in activity against the southern armyworm compared to avermectin B_1 Fig. 16. Interestingly the 22,23-dihydro analogs were more effective than their unsaturated counterparts. However, although a wide variety of monosaccharides and aglycones were synthesized and tested the EC_{90} could not be improved over 0.5 ppm.

Fig. 16

ACTIVITY OF AVERMECTIN DERIVATIVE

AGAINST

SOUTHERN ARMYWORM **NEONATES ON SIEVA BEANS**

Compound	EC_{90} PPM
Avermectin B_1	8.0
Ivermectin	8.0
Avermectin B_1 Monosaccharide	8.0
Ivermectin Monosaccharide	0.5
Ivermectin Aglycone	> 0.5
13-Deoxy IVM Aglycone	0.5
13-β-Cl-13-deoxy IVM Aglycone	0.5
13-β-F-13-deoxy IVM Aglycone	0.5
13=NOCH$_3$-13-deoxy IVM Aglycone	0.5

An important breakthrough came with the discovery of 4"-aminoavermectins. It was reasoned that since many macrolides contain amino sugars it could be interesting to devise a synthesis of avermectins also containing amino sugars. The synthetic scheme is shown in Fig. 17[6]. Avermectin B1 was protected at the 5-position as a TBDMS derivative and then oxidized under Swern oxidation conditions to provide the 4"-keto derivative. Reductive amination with ammonium acetate and sodium borohydride, followed by deprotection, gave the axial epiamino derivative as the major product, a smaller amount of the equatorial amino analog and smaller amount of 4"-epiavermectin B1. N-alkylated derivatives were synthesized either by reductive amination using alkylamines or by alkylation of 4"-amino-4"-deoxyavermectins.

Fig. 17

SYNTHESIS OF 4"−EPIAMINOAVERMECTINS

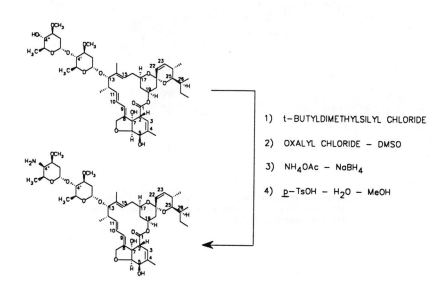

1) t−BUTYLDIMETHYLSILYL CHLORIDE

2) OXALYL CHLORIDE − DMSO

3) NH$_4$OAc − NaBH$_4$

4) p−TsOH − H$_2$O − MeOH

The two epimeric 4"-amino-4"-deoxyavermectin B_1 derivatives had similar biological properties with the 4"-epiamino isomer being a somewhat more potent insecticide. Since the 4"-epiamino derivatives were also the major products of reductive amination, they were selected for further study.

The most active member of the series was 4"-deoxy-4"-epimethylamino avermectin B_1 which has been selected for development as an agricultural insecticide and assigned the code name MK-243 Fig. 18.

Fig. 18

MK-243

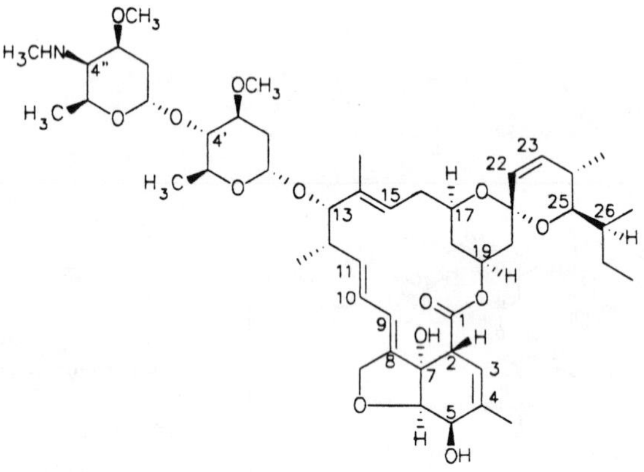

H. Mrozik, P. Eskola, B.O. Linn, A. Lusi, T.L. Shih, M. Tischler, F.S. Waksmunski, M.J. Wyvratt, N.J. Hilton, T.E. Anderson, J.R. Babu, R.A. Dybas, F.A. Preiser and M.H. Fisher, *Experientia*, **45**, 315 (1989).

A summary of the foliar ingestion activity of MK-243 against a variety of insect larvae and adult spider mites and aphids is shown in Fig. 19.

Fig. 19

FOLIAR INGESTION ACTIVITY OF

4''-EPI-METHYLAMINO-4'-DEOXYAVERMECTIN B₁

AGAINST

INSECT LARVAE AND ADULT SPIDER MITES AND APHIDS

SPECIES (Common Name)	LC₉₀(ppm) at 96 hours
Manduca sexta (L.) (tobacco hornworm)	0.003
Trichoplusia ni (Huebner) (cabbage looper)	0.014
Spodoptera exigua (Huebner) (beet armyworm)	0.005
Spodoptera frugiperda (J.E. Smith) (fall armyworm)	0.01
Leptinotarsa decemlineata (Say) (colorado potato beetle)	0.032
Epilachna varivestis (Mulsant) (Mexican bean beetle)	0.20
Tetranychus urticae (Koch) (two-spotted spider mite)	0.29
Aphis fabae (Scopoli) (bean aphid)	19.9

R.A. Dybas, N.J. Hilton, J.R. Babu, F.A. Preiser, and G.J. Dolce, *Proc. Soc. Ind. Microbiol. Int. Conf. Biotech. Microb. Prod.*, San Diego 3/13/88.

References

1.   G. Albers-Schönberg, B. H. Arison, J. C. Chabala, A. W. Douglas, P. Eskola, M. H. Fisher, A. Lusi,H. Mrozik, J. L. Smith and R. L. Tolman, J. Am. Chem. Soc., 1981, 103 4216.

2.   H. Mishima, M. Kurabayashi, C. Tamura, S. Sato, H. Kuwano and A. Saito, Tetrahedron Lett., 1975, 711.

3.   M. V. J. Ramsay, S. M. Roberts, J. C. Russell, A. H. Shingler, A. M. Z. Slawin, D. R. Sutherland, E. P. Tiley and D. J. Williams, Tet. Lett., 1987, 28 , 5353.

4.   G. T. Carter, J. A. Nietsche and D. B. Borders, Chem. Comm., 1987, 402.

5.   S. E. Blanchflower, R. J. J. Dorgan, J. R. Everett and S. A. Readshaw, Tet. Lett., 1988, 29, 6645.

6.   H. Mrozik, P. Eskola, B. O. Linn, A. Lusi, T. L. Shih, M. Tischler, F. S. Waksmunski, M. J. Wyvratt, N. J. Hilton, T. E. Anderson, J. R. Babu, R. A. Dybas, F. A. Preiser and M. H. Fisher, Experientia, 1989, 45, 315.

Synthetic Approaches to Milbemycins

Eric J. Thomas
DEPARTMENT OF CHEMISTRY, THE UNIVERSITY OF MANCHESTER, MANCHESTER M13 9PL, UK

1 INTRODUCTION

The milbemycins and avermectins are attracting considerable interest at present from synthetic organic chemists because of their challenging molecular structures and potent and useful biological activities.[1] Following the syntheses of milbemycin ß3,[2] several syntheses of avermectins have been developed.[3] Our work in this area was initially concerned with a total synthesis of the non-aromatic ß-milbemycins. We now wish to describe a synthesis of milbemycin E (1), together with preliminary results concerned with a projected synthesis of avermectins, e.g. A2a/b (2) and (3).

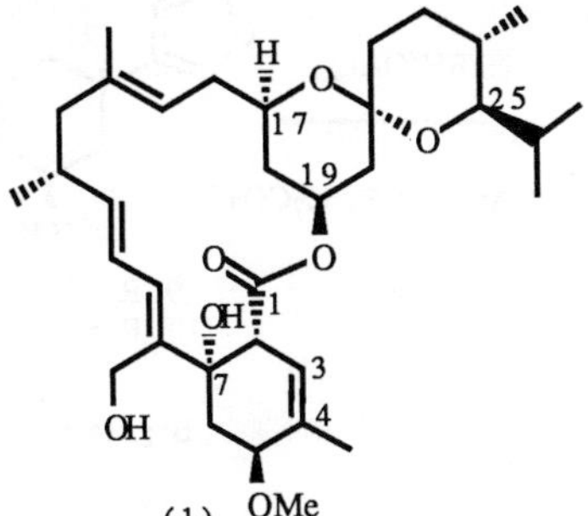

(1)

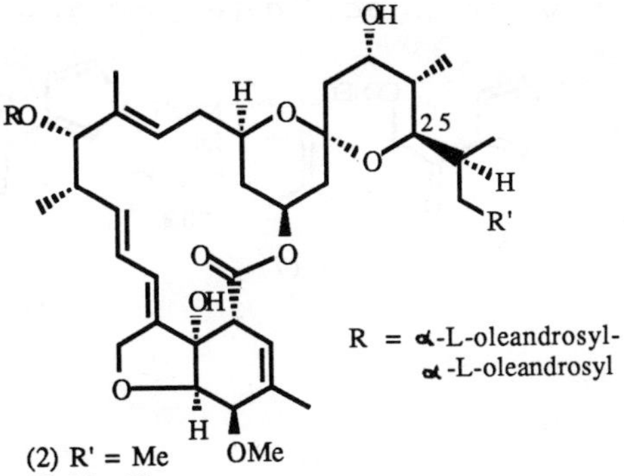

(2) R' = Me
(3) R' = H

R = α-L-oleandrosyl-
    α-L-oleandrosyl

2 MODEL STUDIES AND STRATEGY

At the onset of our work the dihydroxyester (4), which contains some of the features of the C(1)-C(7) fragment of milbemycin E, had been prepared by the Robinson annelation of ethyl benzoylacetate and methyl vinyl ketone followed by stereoselective reduction.[4] It was decided to synthesize the furan containing ester (5) using this approach to see whether the furan ring could be used to introduce the conjugated diene fragment of milbemycin E.[5] Future work would then include introduction of the C(3)-C(4) double-bond, and the development of an asymmetric synthesis.

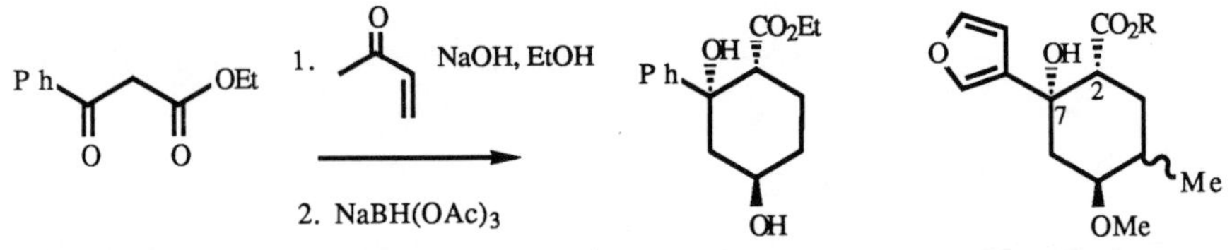

(4)   (5)

Robinson annelation of ketoester (6) and isopropenyl methyl ketone (7) gave adduct (8) isolated as a single diastereoisomer after recrystallization. Reduction using NaBH(OAc)$_3$ gave diol (9) which was converted into the mono-methyl ether (10) using Me$_3$O$^+$ BF$_4^-$ in the presence of K$_2$CO$_3$.

Regioselective modification of the furan ring was discovered serendipitously. Treatment with bromine in methanol gave a mixture of the dihydrodimethoxyfurans (11) which on acid catalysed hydrolysis gave the 3-substituted

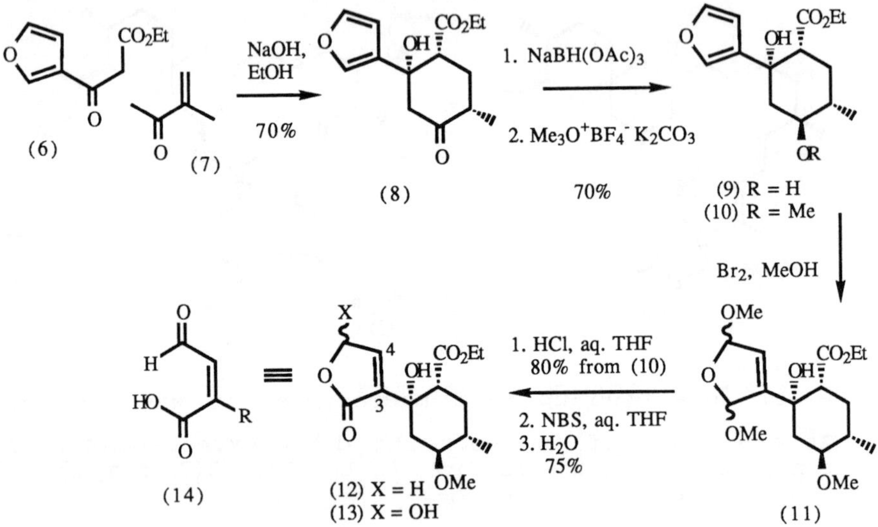

(6)   (7)   (8)   (9) R = H
                                        (10) R = Me

(14)   (12) X = H
              (13) X = OH

(11)

butenolide (12) containing none of the 4-substituted isomer. The origin of this unexpected regioselectivity was not investigated, but may involve participation of the tertiary hydroxyl. Bromination of the butenolide followed by hydrolysis gave the hydroxybutenolide (13) as a mixture of epimers. This hydroxybutenolide is the synthetic equivalent of aldehyde (14), and so its use for conjugated diene synthesis was investigated.

Treatment of hydroxybutenolide (13) with an excess of the simple phosphorane (15) gave a mixture of the Z,Z- and Z,E-dienes (16) and (17) which was converted into the Z,E-isomer (17) containing only a trace of the unwanted Z,Z-isomer (16) using a catalytic quantity of iodine in benzene. However the diene-acid (17) could not be converted into the corresponding alcohol (18). Many procedures for activation of the acid, *e.g.* mixed anhydride formation, were investigated, and a range of reducing agents were used, but a procedure for the efficient formation of alcohol (18) was not found. Either unchanged starting material was recovered, or a complex mixture of products was obtained.

It was thought that the formation of these complex mixtures of products could be due to lactonization of the initially formed dihydroxy-ester (18) followed by dehydration and further decomposition. Since lactonization of the hydroxymethyl group would not be possible if the ester were tied up in a macrocycle, it was decided to investigate the synthesis of a simple macrocyclic analogue of milbemycin E using the hydroxybutenolide (13).

Condensation of the racemic phosphorane (19) with hydroxybutenolide (13) was achieved using two equivalents of

lithium hexamethyldisilazide to deprotonate the
hydroxybutenolide before addition of the ylid. Isomerization
of the product mixture with iodine as before gave a 1 : 1

mixture of the isomeric dienes (20) and (21) which were converted into the hydroxyacids (22) and (23) *via* a three step procedure involving ester exchange using trimethyl-silylethanol and base, esterification using diazomethane, and simultaneous alcohol deprotection and silylethyl ester cleavage using tetrabutylammonium fluoride and acid. Cyclization of this mixture using N-methyl-2-chloro-pyridinium iodide gave a single cyclic product identified as the macrolide (24) by a single crystal X-ray diffraction study. Reduction of (24) using REDAL gave the corresponding alcohol (25) in 80% yield so completing a synthesis of a racemic macrocyclic milbemycin analogue.

This synthesis of (25) established the use of the furan containing Robinson product (8) for the synthesis of milbemycin-like macrolides, and defined a strategy for a proposed milbemycin E synthesis. In particular it was decided to investigate the synthesis and Wittig coupling of the 'upper hemisphere' phosphorane (26) and the 3-cyclo-hexenyl-hydroxybutenolide (27).

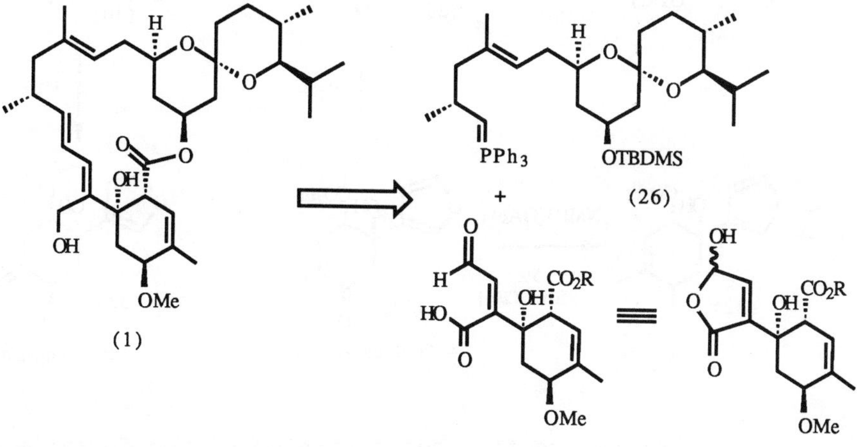

Figure 1 Strategy for a Milbemycin E Synthesis

3 SYNTHESIS OF THE 3-CYCLOHEXENYL-HYDROXYBUTENOLIDE

To synthesize the cyclohexenyl-hydroxybutenolide (27) a procedure was required for the regioselective introduction of the C(3)-C(4) double-bond into the Robinson product (8).[6] Moreover (27) was required as a single enantiomer to avoid formation of mixtures of coupled products corresponding to the diastereoisomers (20) and (21). Attempts to introduce the C(3)-C(4) double-bond by dehydration of a C(4) alcohol led to methylenecyclohexane formation. However conversion of the Robinson product (8) into enol-ether (28) using trimethylsilyltriflate and triethylamine, followed by treatment with benzeneselenenyl chloride and tetrabutyl-ammonium fluoride (to desilylate the tertiary alcohol), gave the phenylselenoketone (29). This on oxidative elimination gave the cyclohexenone (31) containing *ca.* 20% of its exocyclic isomer. These were not separated, but were reduced as a mixture to give the endocyclic and exocyclic alkenes (32) and (33), ratio *ca.* 5 : 1, which were separated by flash chromatography. Interestingly oxidative elimination of the phenylseleno-alcohol (30), prepared by reduction of ketone (29), gave more of the exocyclic alkene product (33).

A new procedure for the conversion of the furan into a hydroxybutenolide was required since the bromine-methanol sequence used earlier for hydroxybutenolide formation was incompatible with the presence of the C(3)-C(4) double-bond. The oxidation of 2-trimethylsilylfurans by $^{1}O_2$ has been used for the regioselective synthesis of hydroxybutenolides,[7] and so the application of this procedure for the synthesis of hydroxybutenolide (27) was investigated together with resolution of the Robinson products.

Robinson annelation of the trimethylsilylfuranyl-keto-ester (34) with methyl isopropenyl ketone (7) gave the hydroxycyclohexanone (35) which was reduced using NaBH(OAc)$_3$

to give the racemic cyclohexanediol (36). Condensation with
the acetate of (<u>S</u>)-(+)-mandelic acid gave a mixture of the
diastereomeric esters (37) and (39), from which one
diastereoisomer crystallized out very efficiently, but was
shown by <u>X</u>-ray diffraction to correspond to the unwanted
isomer (39). The mother liquor from this crystallization was
treated with K_2CO_3 in ethanol to effect selective acetate
cleavage, and from the mixture of alcohols so obtained, the
desired isomer (38) was crystallized out in good yield.
Further treatment with K_2CO_3 in ethanol gave the
dextrorotatory alcohol (36) [35% overall from racemic (36)].
[In future this resolution would be better carried out using
(<u>R</u>)-(-)-mandelic acid which should provide the enantiomer of
(39) and hence the desired enantiomer of (36) directly.]

Monomethylation of the (+)-diol (36) followed by ester
hydrolysis and re-esterification with trimethylsilylethanol
gave the TMS-ethyl ester (42). Oxidation using $^1O_2$ then gave
the crystalline hydroxybutenolide (43) as a mixture of
epimers. The cyclohexenylhydroxybutenolide (27) was prepared
from the (+)-alcohol (36) by Swern oxidation followed by
phenylselenylation to give the phenylselenoketone (44).
Oxidative elimination, ketone reduction, monomethylation and
ester exchange then gave the cyclohexenylfuran (45) which

was converted efficiently into the hydroxybutenolide (27) using $^1O_2$. The hydroxybutenolide (27) so obtained was a mixture of epimers, but was a highly crystalline solid, and was found to undergo Wittig reactions with simple phosphoranes.

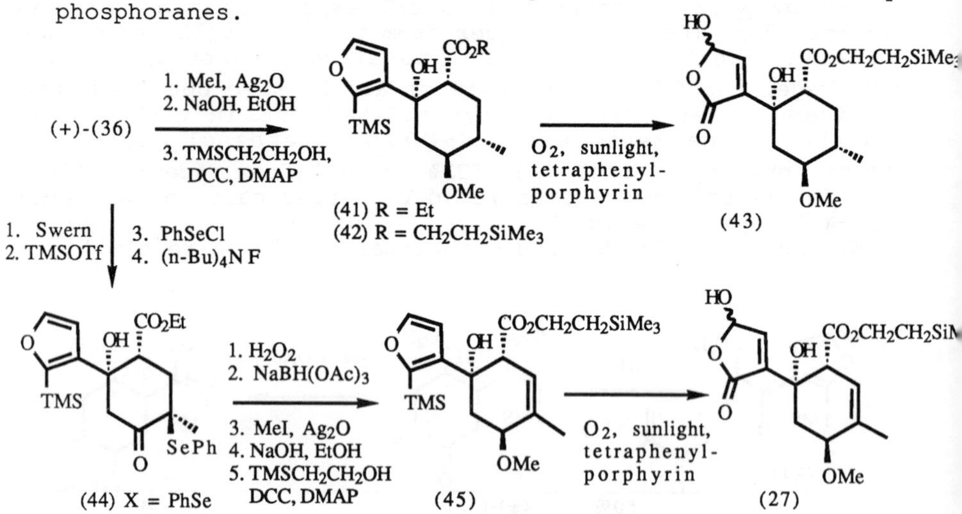

4 SYNTHESIS OF THE 'UPPER HEMISPHERE' PHOSPHORANE

It was proposed to assemble the open chain equivalent (46) of the spiroacetal of milbemycin E by alkylation of 1,3-dithiane using iodide (48) and epoxide (47) as shown in Figure 2. Deprotection would then lead to the spiroacetal.[8]

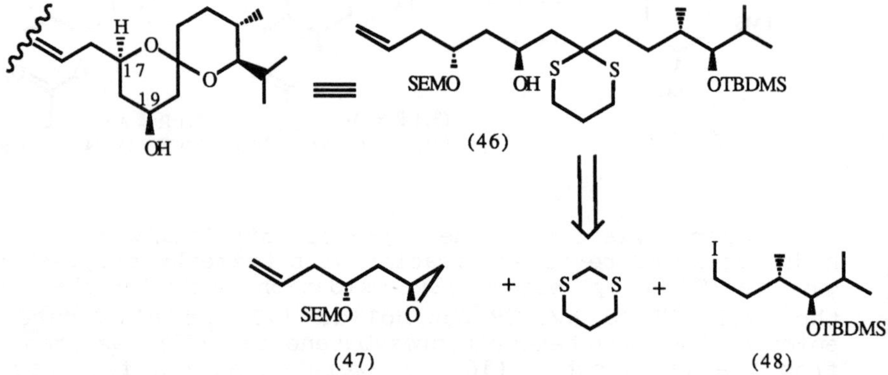

Figure 2 Strategy for Milbemycin E Spiroacetal Synthesis

A short enantioselective synthesis of the iodide (48) was developed based on H. C. Brown's diisopinocampheylborane chemistry. Treatment of 2-methylpropanal with the crotyldiisopinocampheylborane (49) prepared from $\underline{E}$-but-2-ene and (+)-α-pinene gave a mixture of the *anti-* and *syn-*hydroxyalkenes (50) and (51), ratio 85:15. From its Mosher's derivative the optical purity of the major adduct (50) was estimated to correspond to an e.e. of 90%, and oxidative cleavage, followed by reduction, gave diol (52) which was dextrorotatory, so confirming the absolute configuration as shown. Protection of the alcohol followed by regioselective hydroboration and treatment with iodine, triphenylphosphine, and imidazole, gave the iodide (48).

The protected hydroxy-epoxide (47) was prepared from ($\underline{S}$)-malic acid. Reduction of dimethyl ($\underline{S}$)-malate using borane-dimethylsulphide complex[10] and a catalytic amount of sodium borohydride gave diol (55) which was converted into acetal-aldehyde (57) *via* diol protection, LiAlH$_4$ reduction, and Swern oxidation. Treatment of this aldehyde with simple organometallic reagents does not generally proceed with effective diastereoface selectivity. However use of the allyldiisopinocampheylborane (58) derived from (-)-α-pinene gave a 9:1 mixture of the alcohols (58) and (59). Protection of alcohol (58) with SEM-chloride, followed by acetal hydrolysis and epoxide formation using diethyl azodicarboxylate and triphenylphosphine gave the required epoxyalcohol (47). The stereochemistry of (47) was confirmed by conversion into the acetal (61). This was obtained as a single diastereoisomer, and its [1]H n.m.r. spectrum showed clearly the non-equivalence of H(4) and H(6).

To assemble the spiroacetal, 1,3-dithiane was alkylated using n-butyl-lithium and iodide (48). The second dithiane alkylation was then achieved using t-butyl-lithium and epoxide (47), and the product deprotected to give the spiroacetal (64) directly. The structure assigned to spiroacetal (64) was consistent with its spectroscopic data, including an n.O.e. enhancement of H(17) on irradiation of H(25), and *vice versa*, and the spiroacetal was identical with another sample prepared independently from α-methyl-D-glucoside.[11]

 Spiroacetal (64) was converted into the 'upper hemisphere' of milbemycin E (1) following the approach developed by Baker.[12] Alcohol protection, ozonolysis with a dimethyl sulphide work-up, and condensation with Ph$_3$P=C.CH$_3$.CO$_2$Et, gave ester (65), which was converted into the iodide (66) in two steps. This iodide was used to alkylate the lithium enolate of the Evan's reagent (67) and the product (68) reduced to the alcohol (69). Conversion to the phosphonium salt (71) was achieved *via* the iodide (70).

5 SYNTHESIS OF MILBEMYCIN E

To develop conditions for the Wittig coupling and cyclization steps, the synthesis of 3,4-dihydromilbemycin E (75) was investigated first. Generation of the ylid (26) from phosphonium salt (71) was not straightforward; under the usual conditions for ylid formation, i.e. n-butyl-lithium at 0 °C, the ylid was found to be unstable, and only low yields of product were obtained. However using t-butyl-lithium at -40 °C clean ylid formation occurred. The Wittig reaction with hydroxybutenolide (43) was then carried out by adding a solution of ylid (26) containing two mole equivalents of lithium hexamethyldisilazide to a solution of the hydroxybutenolide at -78 °C, and stirring the reaction mixture for several hours at -40 °C during which time the red colour of the ylid disappeared. The crude product from the reaction was esterified with diazomethane and isomerized with a trace of iodine to give the coupled product (72), 60% overall from the phosphonium salt (71), identified from spectroscopic data by comparison with the other Wittig products prepared earlier. Less than 5% of the 10,11-Z-isomer was detected by ^{1}H n.m.r.

Deprotection of the Wittig product (72) using tetra-butylammonium fluoride gave hydroxyester (73) which was cyclized at 0 °C using dicyclohexylcarbodiimide and 4-dimethylaminopyridine to give the macrolide (74). Finally reduction of the methyl ester group using REDAL gave 3,4-dihydromilbemycin E (75), which was identified on the basis of spectroscopic data, and by comparison with milbemycin E (1), *vide infra*.

The procedures developed for the synthesis of dihydromilbemycin E (75) were then applied to synthesize milbemycin E itself. Addition of the ylid (26) and lithium hexamethyldisilazide to the cyclohexenylhydroxybutenolide

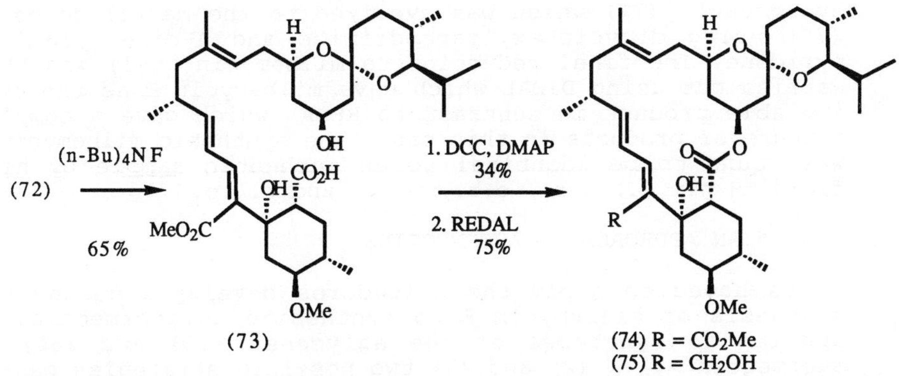

(72) → (73)

(72) 65%

(73)

1. DCC, DMAP 34%

2. REDAL 75%

(74) R = CO₂Me
(75) R = CH₂OH

(74) R = CO_2Me
(75) R = CH_2OH

(27) gave the coupled ester (76) after esterification with diazomethane and iodine isomerization. As before the iodine catalysed isomerization was very effective, only a trace, less than 2%, of the unwanted 10,11-Z-isomer could be detected by high field ¹H n.m.r. Deprotection gave the

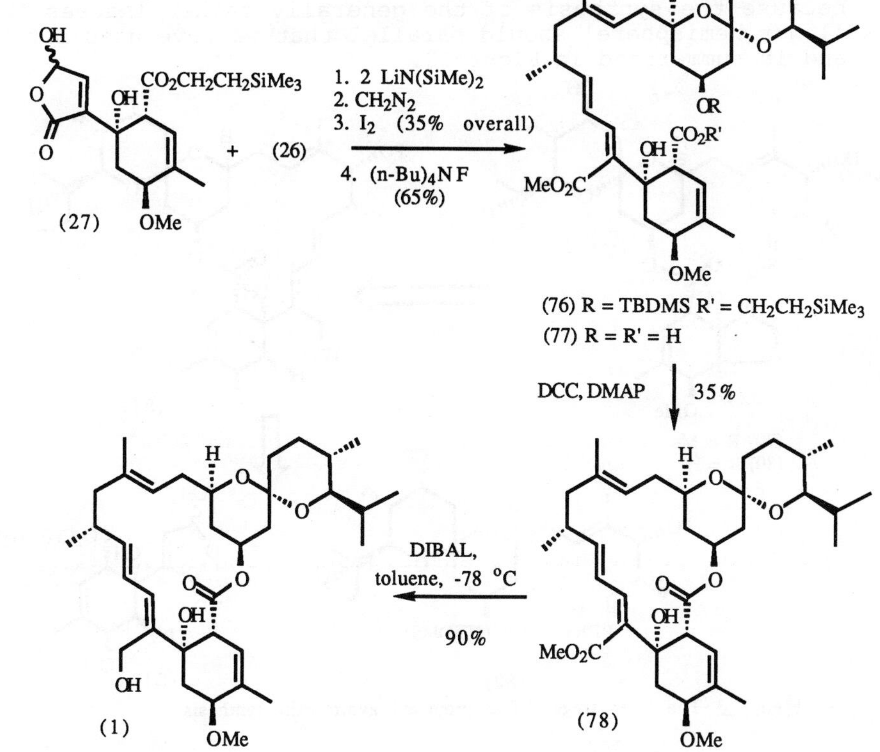

(27)

+ (26)

1. 2 LiN(SiMe)₂
2. CH₂N₂
3. I₂ (35% overall)

4. (n-Bu)₄NF (65%)

(76) R = TBDMS R' = CH₂CH₂SiMe₃
(77) R = R' = H

DCC, DMAP   35%

DIBAL, toluene, -78 °C

90%

(1)

(78)

hydroxyacid (77) which was cyclized to the macrolide ester
(78) using dicyclohexylcarbodiimide and 4-dimethylamino-
pyridine. The final reduction to milbemycin E (1) was then
carried out using DIBAL which gave milbemycin E as the only
isolable product in contrast to REDAL which gave a complex
mixture of products in this case. The synthetic milbemycin E
was found to be identical to an authentic sample by high
field ^{1}H n.m.r., i.r., o.r., m.s., and t.l.c.

6 AN APPROACH TO AVERMECTINS

It is hoped to apply the procedures developed during the
synthesis of milbemycin E to synthesize an avermectin. As
applied to syntheses of the aglycones (79) and (80) of
avermectin $A_{2a/b}$ (2) and (3) two possible strategies can be
considered. Either a 'lower hemisphere' aldehyde containing
the required bicyclic nucleus could be synthesized, or a
hydroxybutenolide analogous to that used in the milbemycin E
synthesis could be used with the formation of the
tetrahydrofuran ring being left until the end of the
synthesis. This latter 'high risk' strategy is attractive
because the synthesis of the generally rather inaccessible
'lower hemisphere' should parallel that we have used before,
and is summarized in Figure 3.

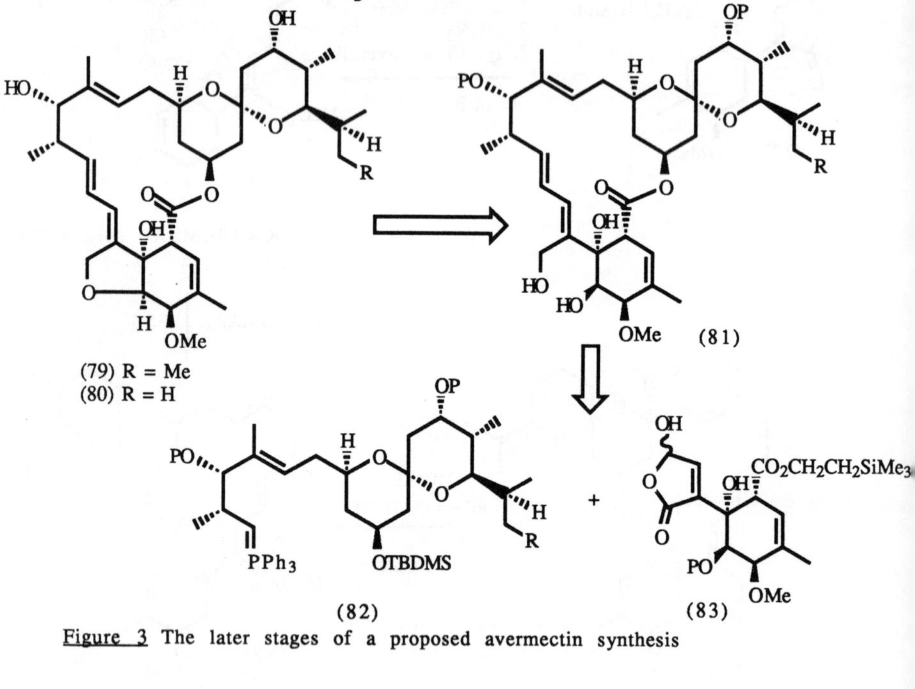

(79) R = Me
(80) R = H

(81)

(82)   +   (83)

Figure 3 The later stages of a proposed avermectin synthesis

Considerable progress has been made with this synthesis, in particular an alcohol corresponding to the 'upper hemisphere' ylid has been synthesized, and Robinson annelation products required for the synthesis of the hydroxybutenolide (83) have been prepared.

Figure 4 outlines the strategy used in the avermectin spiroacetal synthesis. This parallels that used for the synthesis of the milbemycin spiroacetal (64), but required the synthesis of the new hydroxyepoxides (88) and (89).

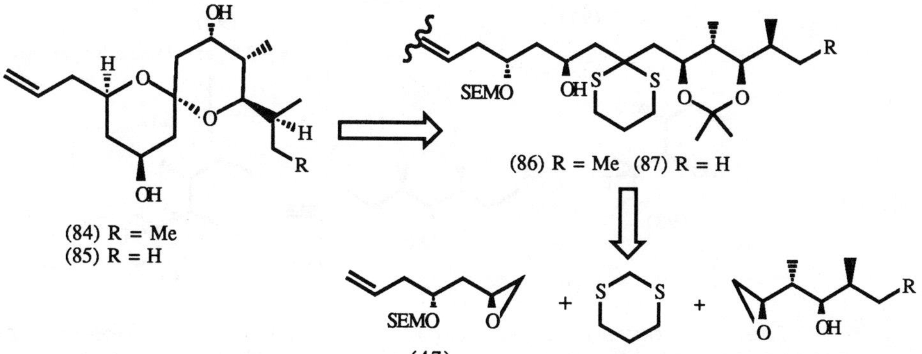

(84) R = Me
(85) R = H

(86) R = Me (87) R = H

(47)

(88) R = Me
(89) R = H

Figure 4.

The starting material for the synthesis of the hydroxyepoxide (88) was (S)-2-methylbutanal (91) prepared by oxidation of the commercially available alcohol (90) using chromium trioxide. In our hands this oxidation was accompanied by some racemization and gave (S)-2-methylbutanal (91) with an e.e. of only *ca.* 70%. However treatment with the crotyldiisopinocampheylborane (49) prepared from E-but-2-ene and (+)-pinene, followed by hydrogen peroxide oxidation, gave a mixture of products from which the required alcohol (92) could be isolated in a reasonable yield [75% of the mixture, 45% of adduct (92)]. Interestingly the optical purity of alcohol (92) was found to correspond to an e.e. of greater than 90%, this increase in optical purity relative to the starting aldehyde being ascribed to the preferential formation of alcohol (93) from (R)-2-methylbutanal and borane (49).

Heptenol (92) was identified by analogy with the results of H. C. Brown, and our synthesis of alcohol (50). Moreover hydrogenation of (92) gave a 3,5-dimethylheptan-

4-ol which was identified as (94) since the two methyl groups were not equivalent in its ^{1}H n.m.r. spectrum. [The absolute configuration of (94) follows from its synthesis from the (S)-aldehyde (91)]

> 98% ee (Mosher's derivative)

The hydroxyepoxides (88) and (89) were prepared by epoxidation of the unsaturated alcohols (92) and (50) using t-butylhydroperoxide in the presence of VO(acac)$_2$. From the oxidation of alcohol (92) two epoxides were isolated and identified by analogy with the literature as the required epoxide (88) (72%) and its diastereoisomer (95) (14%). Only the major epoxide (89) was characterized from the epoxidation of alcohol (50).

Addition of epoxide (88) to an excess of lithiated dithiane gave diol (96) which was protected as its acetonide (97). Deprotonation using t-butyl-lithium followed by the addition of the protected hydroxyepoxide (47) then gave the protected open-chain tetraol (98) which was converted into the spiroacetal (84) in one step using HF.pyridine in aqueous acetonitrile. Hydroxyepoxide (89) was similarly converted into the spiroacetal (85).

(84) R = Me
(85) R = H

Treatment of the spiroacetal (85) with TBDMSOTf protected both the equatorial and axial alcohols, and ozonolysis followed by condensation with $Ph_3P=C.CH_3.CO_2Et$ gave the the bis-protected ester (99). Reduction of the ester with DIBAL followed by a Swern oxidation gave the aldehyde (100) which was reacted with the crotyldi-isopinocampheylborane (49) derived from E-but-2-ene and (+)-pinene to give alcohol (101) after oxidative work-up. Protection using SEM-Cl and selective cleavage of the terminal double-bond then gave the 'upper hemisphere' alcohol (103).

Preliminary studies have been carried out to see whether the Robinson annelation route is suitable for the synthesis of the hydroxybutenolide (83) which has an additional oxygen substituent at C(6) (avermectin numbering).

The efficiency and stereoselectivity of the Robinson annelation reaction between the furanylketo-ester (6) and the benzyloxymethyl isopropenyl ketone (104) was found to depend upon the solvent. In a 1 : 1 mixture of ethanol and

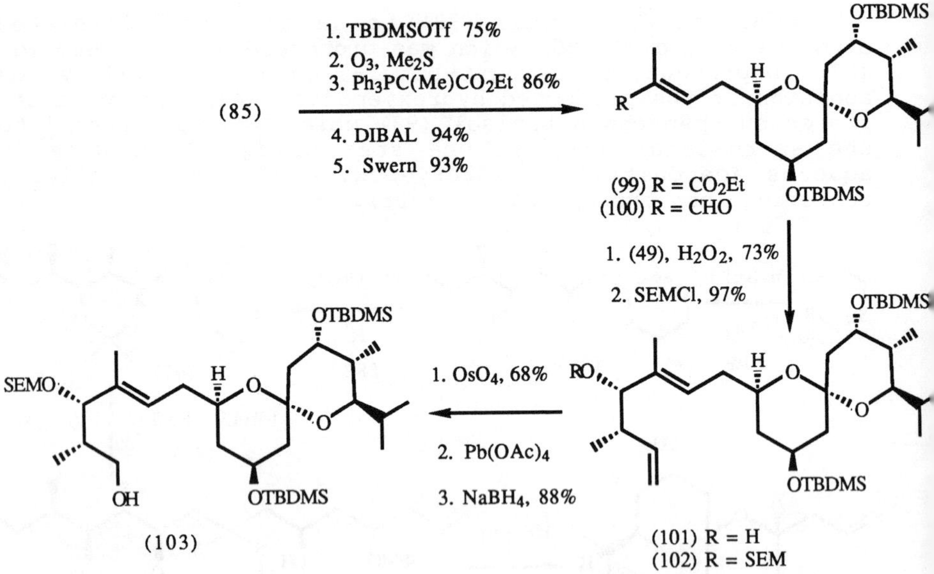

(85)
1. TBDMSOTf 75%
2. O_3, Me_2S
3. $Ph_3PC(Me)CO_2Et$ 86%
4. DIBAL 94%
5. Swern 93%
OTBDMS
H
O
R
O
OTBDMS
(99) R = CO_2Et
(100) R = CHO
1. (49), H_2O_2, 73%
2. SEMCl, 97%
RO
H
O
O
OTBDMS
OTBDMS
(101) R = H
(102) R = SEM
1. OsO_4, 68%
2. $Pb(OAc)_4$
3. $NaBH_4$, 88%
SEMO
H
O
O
OTBDMS
OTBDMS
OH
(103)

dichloromethane, a mixture of three products, which were separated and identified as the adducts (105) - (107), was obtained, the isolated yields of (105), (106), and (107) under these conditions being 20, 15, and 50%, respectively. In contrast, in neat ethanol the adduct (107) crystallized out of the reaction mixture and was isolated in yields of up to 85%, with only a small amount of the epimeric product (106) being left in solution. Similarly in ethanol the adduct (108) was formed as the major product (70%) from the reaction between the trimethylsilylfuranylketo-ester (34) and the benzyloxymethylketone (104).

O
R
CO_2Et
O
(6) R = H
(34) R = TMS
$PhCH_2O$
O
(104)
NaOH
O
OH
CO_2Et
$PhCH_2O$
O
(105)
+
O
OH
CO_2Et
$PhCH_2O$
O
(106)
+
O
R
OH
CO_2Et
2
7
6
$PhCH_2O$
O
(107) R =H
(108) R = TMS

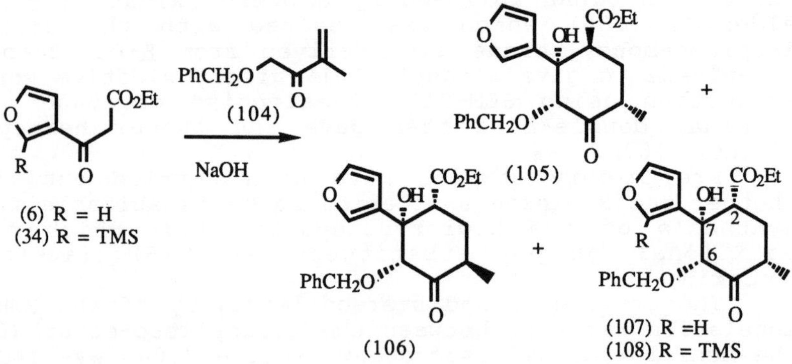

The structures of the Robinson products (105) - (108) were established on the basis of n.m.r. studies including n.o.e. experiments, and their chemistry was similar to that of analogous compounds prepared earlier. For example reduction of adduct (107) using $NaBH_4$ gave diol (109) whereas reduction using $NaBH(OAc)_3$ gave diol (110).

| (109) | NaBH₄ ← | (107) | → NaBH(OAc)₃ | (110) |

The major adducts (107) and (108) both have the required stereochemistry at C(2) and C(7) for conversion into the target hydroxybutenolide (83), but their configurations at C(6) will need to be inverted. Nevertheless the accessability of these compounds makes them attractive intermediates for an avermectin synthesis. Present work is concerned with the conversion of adduct (108) into the hydroxybutenolide (83) and with the preparation of the bicyclic C(1) - C(10) fragment of avermectin A_{2a} (2) from the adduct (107). The condensation of these fragments with the 'upper hemisphere' phosphorane derived from alcohol (103) will then be investigated.

ACKNOWLEDGEMENTS

I should like to thank all my collaborators who have been involved with this work. These include M. J. Hughes (model macrolide), N. A. Stacey and S. V. Mortlock [introduction of the C(3)-C(4) double-bond], P. G. Steel and E. R. Parmee (milbemycin E synthesis), E. Merifield (spiroacetal synthesis), M. Smallridge (avermectin 'upper hemisphere'), and S. Karim (present work on the avermectin 'lower hemisphere'). In addition S. Vather, G. C. Robinson, N. A. Stacey, and G. Khandekar were involved with an alternative synthesis of the milbemycin E spiroacetal. I should also like to thank Dr. M. D. Turnbull (I.C.I. Agrochemicals) for many helpful discussions, the S.E.R.C., I.C.I. Agrochemicals, and Rhone-Poulenc for their support, Mr. O. S. Mills (Manchester University) for the X-ray structure determination of ester (39), and Dr. J. Ide (Sankyo Company Limited, Tokyo) for a sample of authentic milbemycin E.

REFERENCES

1. H. G. Davies and R. H. Green, *Nat. Prod. Rep.*, 1986, **3**, 87.

2. A. B. Smith, III, S. R. Schow, J. D. Bloom, A. S. Thompson, and K. N. Winzenberg, *J. Am. Chem. Soc.*, 1982, **104**, 4015; D. R. Williams, B. A. Barner, K. Nishitani, and J. G. Phillips, *ibid.*, p. 4708; R. Baker, M. J. O'Mahony, and C. J. Swain, *J. Chem. Soc., Chem. Commun.*, 1985, 1326; S. D. A. Street, C. Yeates, P. Kocienski, and S. F. Campbell, *ibid.*, p. 1386; C. Yeates, S. D. A. Street, P. Kocienski, and S. F. Campbell, *ibid.*, p. 1388; S. V. Attwood, A. G. M. Barrett, R. A. E. Carr, and G. Richardson, *ibid.*, 1986, p. 479; A. G. M. Barrett, R. A. E. Carr, S. V. Attwood, G. Richardson, and N. D. A. Walshe, *J. Org. Chem.*, 1986, **51**, 4840; S. R. Schow, J. D. Bloom, A. S. Thompson, K. N. Winzenberg, and A. B. Smith, III, *J. Am. Chem. Soc.*, 1986, **108**, 2662; R. Baker, M. J. O'Mahony and C. J. Swain, *J. Chem. Soc., Perkin Trans. I*, 1987, 1623; P. J. Kocienski, S. D. A. Street, C. Yeates, and S. F. Campbell, *ibid.*, p. 2171, 2189; P. J. Kocienski, C. Yeates, S. D. A. Street, and S. F. Campbell, *ibid.*, p. 2183; M. T. Crimmins, D. M. Bankaitis-Davis, and W. G. Hollis, Jr., *J. Org. Chem.*, 1988, **53**, 652.

3. S. Hanessian, A. Ugolini, D. Dube, P. J. Hodges, and C. Andre, *J. Am. Chem. Soc.*, 1986, **108**, 2776; S. Hanessian, A. Ugolini, P. J. Hodges, P. Beaulieu, D. Dube, and C. Andre, *Pure Appl. Chem.*, 1987, **59**, 299; S. J. Danishefsky, D. M. Armistead, F. E. Wincott, H. G. Selnick, and R. Hungate, *J. Am. Chem. Soc.*, 1987, **109**, 8117; S. J. Danishefsky, H. G. Selnick, D. M. Armistead, and F. E. Wincott, *ibid.*, p. 8119.

4. M. D. Turnbull, G. Hatter, and D. E. Ledgerwood, *Tetrahedron Lett.*, 1984, **25**, 5449.

5. M. J. Hughes, E. J. Thomas, M. D. Turnbull, R. H. Jones, and R. E. Warner, *J. Chem. Soc., Chem. Commun.*, 1985, 755.

6. S. V. Mortlock, N. A. Stacey, and E. J. Thomas, *J. Chem. Soc., Chem. Commun.*, 1987, 880.

7. S. Katsumura, K. Hori, S. Fugiwara, and S. Isoe, *Tetrahedron Lett.*, 1985, **26**, 4625.

8. E. Merifield, P. G. Steel, and E. J. Thomas, *J. Chem. Soc., Chem. Commun.*, 1987, 1826.

9. H. C. Brown and K. S. Bhat, *J. Am. Chem. Soc.*, 1986, **108**, 5919.

10. S. Saito, T. Hasegawa, M. Inaba, R. Nishida, T. Fujii, S. Nomizu, and T. Moriwake, *Chemistry Lett.*, 1984, 1389.

11. G. Khandekar, G. C. Robinson, N. A. Stacey, P. G. Steel, E. J. Thomas, and S. Vather, *J. Chem. Soc., Chem. Commun.*, 1987, 877.

12. R. Baker, M. J. O'Mahoney, and C. J. Swain, *Tetrahedron Lett.*, 1986, **27**, 3059.

Synthesis and Modification of Azadirachtin and Related Antifeedants

Steven V. Ley
DEPARTMENT OF CHEMISTRY, IMPERIAL COLLEGE OF SCIENCE, TECHNOLOGY AND
MEDICINE, LONDON SW7 2AY, UK

The need to protect our food supply from predatory
insect attack using more ecologically acceptable methods
has led to a rapidly growing interest in behaviour
modifying chemicals from natural sources. Insect
antifeedants have attracted particular attention owing to
their potential for species selectivity and incorporation
into pest integrated management schemes. Current
environmental pressures and rapidly developing resistance
to conventional insecticides provides the impetus for
future research.

We became interested in the synthesis of
antifeedants several years ago. Our aim was to devise
routes to these natural products and to determine the
functional groups responsible for their biological
activity. Ultimately, with careful entomological
evaluation, this research may well provide a basis for
the design of structurally less complex antifeedants
capable of mimicking the natural product and amenable to
commercial development. In our early work we developed
highly efficient routes to the drimane antifeedants
polygodial and warburganal.[1]

Polygodial   Warburganal

We also evaluated structural analogues of these compounds which displayed antifeedant effects against a variety of crop pests.[2] Further synthetic studies on the more structurally complex clerodane diterpenoid antifeedants culminated in the first total synthesis[3] of ajugarin I, a compound isolated from the *Ajuga remota* plant. We characterised the key functional groups in this molecule which were essential for its activity[4] and observed that further functional changes in the side chain produced notable differences in the antifeedant effect against various insect species.[5]

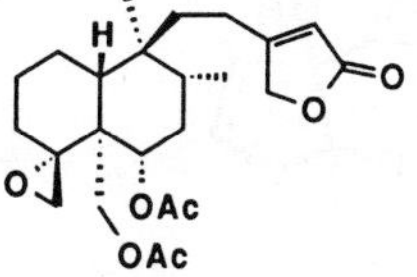

Ajugarin 1

More recently, with our colleagues at Kew, we have been able to isolate and characterise a number of new clerodane compounds from the genus *Scuttelaria*.[6] The structures of these new compounds are shown in Scheme 1 together with some initial bioassay results using the Egyptian Cotton leaf worm *Spodoptera littoralis*. Noticeable from the table is that the compound jodrellin B is some 5-10 times more active than other common clerodane antifeedants such as clerodin. This exciting result provides further useful structural characteristics potentially important for activity or in the design of novel analogue compounds.

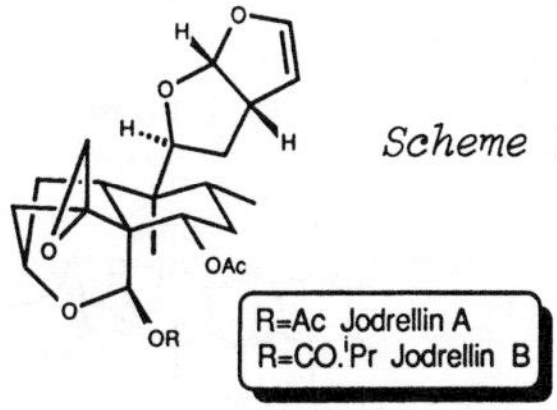

Scheme 1

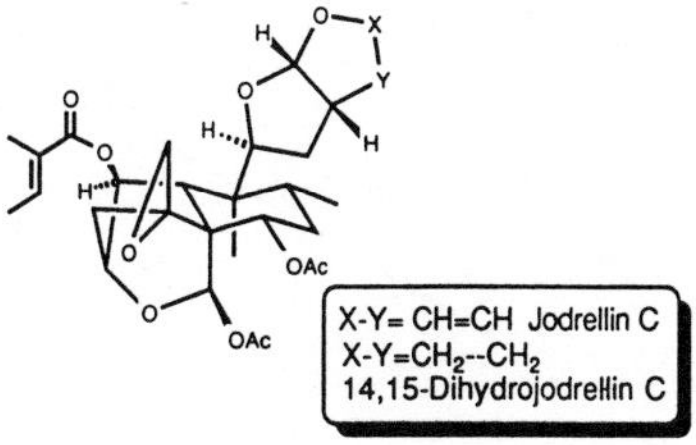

R=Ac Jodrellin A
R=CO.^{i}Pr Jodrellin B

X-Y= CH=CH Jodrellin C
X-Y=CH₂--CH₂
14,15-Dihydrojodrellin C

Bioassay Results. Antifeedant Index [(C-T)/(C+T)]%

Concentration applied (ppm)	jodrellin B	jodrellin A	clerodin
100	100 ± 0.0	92 ± 7.6	74 ± 8.4
25	83 ± 10.3	53 ± 13.3	24 ± 9.8
1	54 ± 14.4	43 ± 15.9	14 ± 19.6

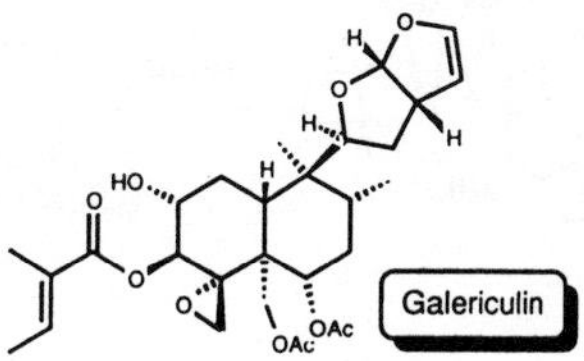

Galericulin

The neem tree *Azadirachta indica* has been known for centuries as a source of useful materials. However it was not until in 1968, when Morgan succeeded in isolating, in pure form, a compound he called azadirachtin, that real scientific work began. This showed extremely potent antifeedant and growth disruption properties. The structure of azadirachtin has been the subject of much debate. It was not until 1985 that our group and the Kraus group, in Hoenheim, working independently, finally assigned the correct structure.[7]

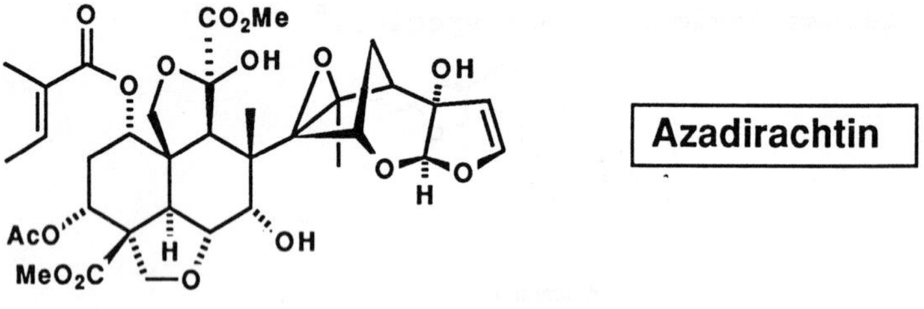

Azadirachtin

In this work we were able to use X-ray crystallographic techniques to unambiguously assign all the structural features of this novel molecule. Furthermore these studies provided chemically modified azadirachtins which are important in the determination of structure activity profiles (Scheme 2)

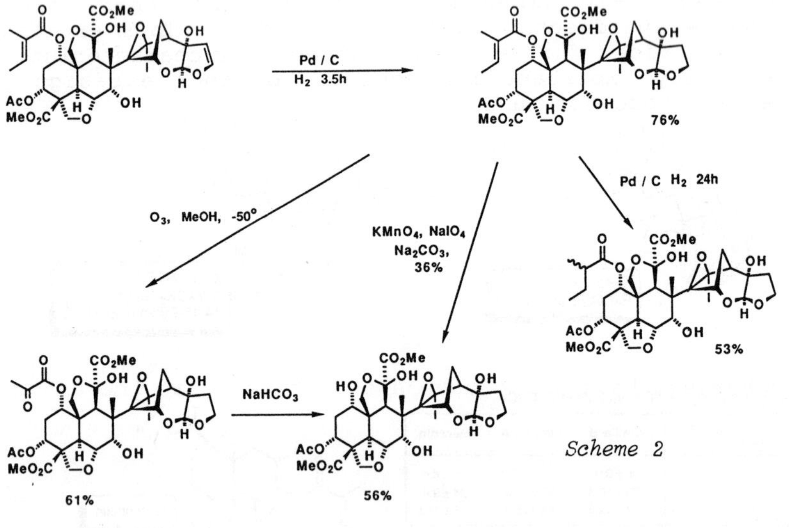

Scheme 2

Many other chemical modifications of azadirachtin have now been accomplished by our group some of which are reported below. A full account of this work and the related biological evaluations will appear shortly.[8]

Azadirachtin readily undergoes addition of bromine and alcohols to the 22,23 double bond and even addition of acetic acid proceeds smoothly (Scheme 3).[9]

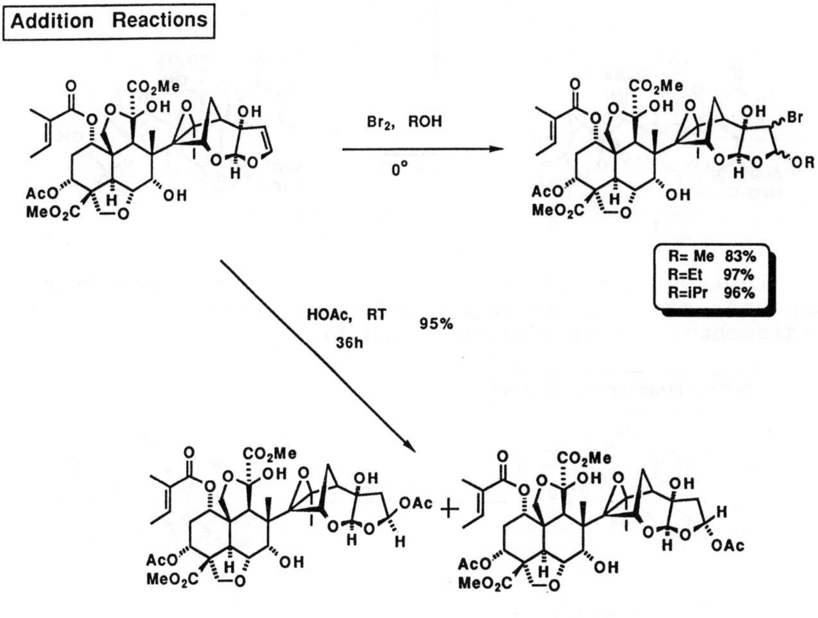

Scheme 3

The products of these reactions are useful for further synthesis. We have also noticed that the dihydro azadirachtin skeleton may be further rearranged to give azadirachtinins upon exposure to acid catalysis which effects epoxide ring opening at C-13 by the hindered C-7 hydroxyl group.[10] (Scheme 4).

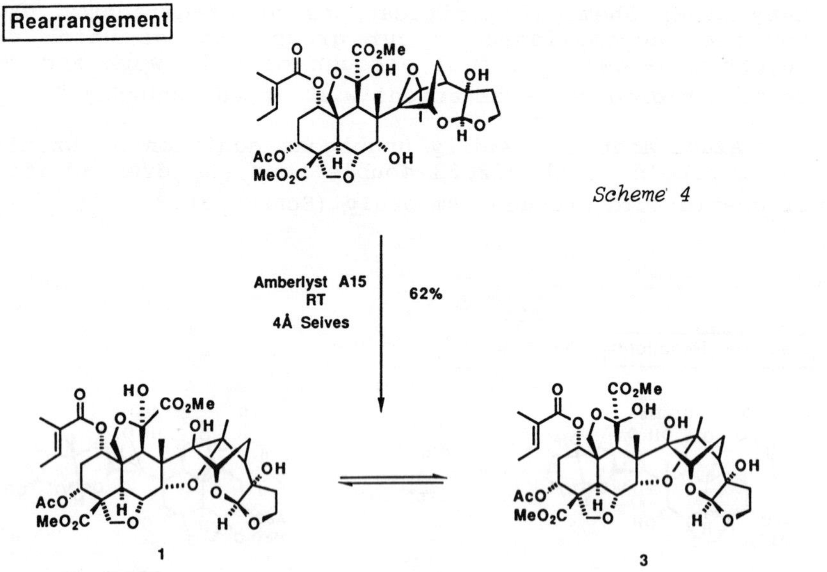

Rearrangement

Scheme 4

Amberlyst A15
RT   62%
4Å Seives

1   ⇌   3

Much of the chemistry common to the azadirachtin compounds may also be applied to the related 3-tigloyl-azadirachtol series (Scheme 5 and 6)

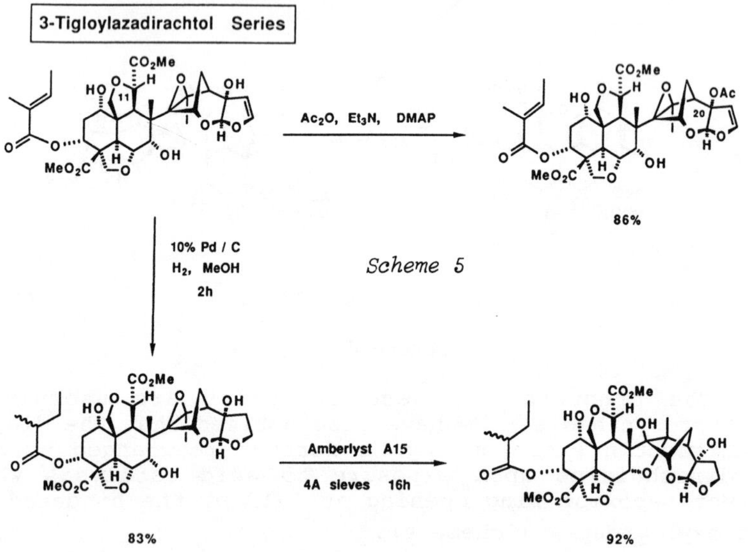

3-Tigloylazadirachtol Series

Ac$_2$O, Et$_3$N, DMAP

86%

10% Pd / C
H$_2$, MeOH
2h

Scheme 5

83%

Amberlyst A15

4A sieves   16h

92%

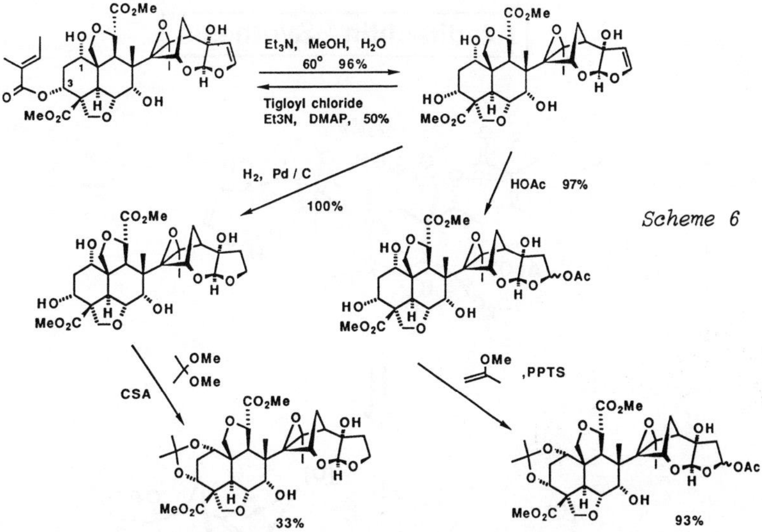

Scheme 6

In addition to the structural modification programme we have also been pursuing the total synthesis of azadirachtin. This fascinating compound containing 16 asymmetric carbon atoms and a plethora of oxygen functionality presents a formidable challenge to the synthesis chemist. Our strategy to this molecule is described in broad outline in Scheme 7. Essentially we propose to couple a decalin fragment to a tricyclic furan unit. Further structural elaborations to the natural product are fairly obvious but necessitate careful planning and execution.

We have recently reported the preparation of the key decalin fragment necessary for the total synthesis.[11] This involved a number of interesting synthetic steps which have been discussed elsewhere[11] and hence are not discussed in detail in this report.

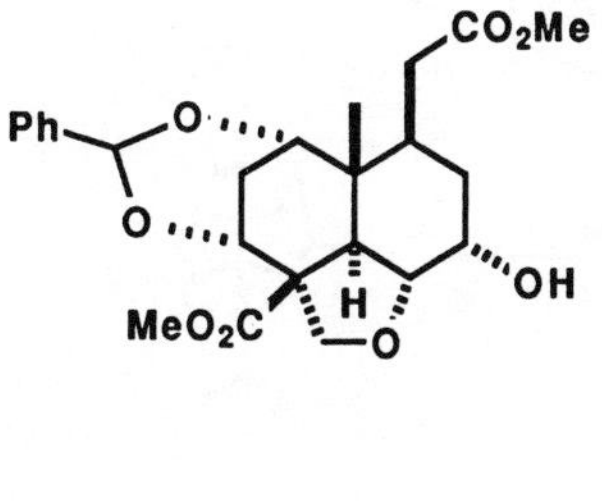

Decalin Fragment

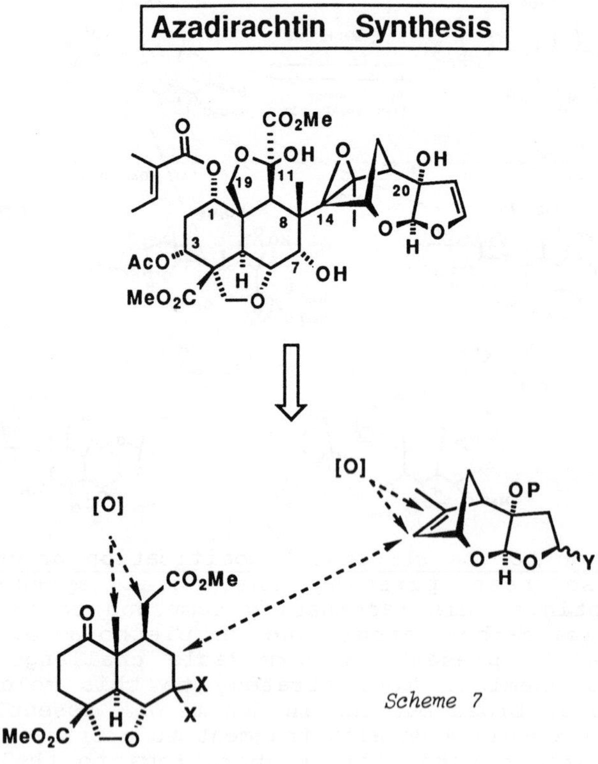

Azadirachtin   Synthesis

Scheme 7

Finally in order to become familiar with and to define a strategy to the tricyclic furan unit, we have synthesised a model system (Scheme 8).

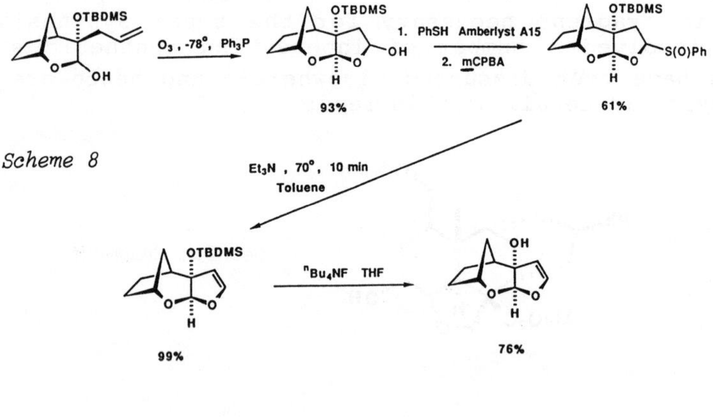

Scheme 8

Additionally this work has provided several novel structures for biological screening. Particularly interesting is that the model fragment similar to the azadirachtin behaves as a very potent antifeedant against *Spodoptera littoralis*.

We believe this work is encouraging for the total synthesis of azadirachtin and especially interesting for the study of antifeedant effects at the molecular level.

ACKNOWLEDGEMENTS:

I thank M. Mahon, R.B. Katz, N.S. Simpkins, A.J. Whittle, W.P. Jackson, D. Craig, A. Abad, P.S. Jones, H.B. Broughton, D. Santafianos, N.G. Robinson, P.L. Toogood, J.C. Anderson and S.C. Smith for the excellent contributions and hard work towards these studies. Also Zev Lidert for valuable discussions, David Morgan for natural product supply, Wally Blaney and Monique Simmonds for biological screening. Finally I acknowledge support from the SERC, ICI Agrochemicals Division, British Technology Group and Rohm and Haas.

REFERENCES:

1. D. M. Hollinshead, S.C. Howell, S.V. Ley, M. Mahon, M.N. Ratchliffe and P.A. Worthington, *J. Chem.Soc. Perkin Trans*. I, 1983, 1579.
2. W.M. Blaney, M.S.J. Simmonds, S.V. Ley and R.B. Katz, *Physiological Entomol.*, 1987, *12*, 281.
3. P.S. Jones, S.V. Ley, N.S. Simpkins and A.J. Whittle, *Tetrahedron*, 1986, *42*, 6519.
4. S.V. Ley, D. Neuhaus, N.S. Simpkins and A.J. Whittle, *J.Chem.Soc. Perkin Trans*. I, 1982, 2157.
5. W.M. Blaney, M.S.J. Simmonds, S.V. Ley, P.S. Jones, *Entomol. Exp. Appl.*, 1988, *46*, 267.
6. J.C. Anderson, W.M. Blaney, M.D. Cole, L.L. Fellows, S.,V. Ley, R.N. Sheppard and M.S.J. Simmonds, *Tetrahedron Lett.*, 1989, *29*, in press.
7. For the full papers see: D.A.H. Taylor, *Tetrahedron*, 1987, *43*, 2779; C.J. Turner, M.S. Tempester, R.B. Taylor, M.G. Zagorski, J.S. Termini, D.R. Schroeder, and K. Nakanishi, *Tetrahedron*, 1987, *43*, 2789; J.N. Bilton, H.B. Broughton, P.S. Jones, S.V. Ley, Z. Lidert, E.D. Morgan, H.S. Rzepa, R.N.Sheppard, A.M.Z. Slawin and D.J. Williams, *Tetrahedron*, 1987, *43*, 2805; W.Kraus, M. Bokel, A. Bruhn, R. Cramer, I. Klaber, A. Klenk, G. Nagl, H. Pöhnl, H. Sadlo and B.

Volger, *Tetrahedron*, 1987, *43*, 2817.
8. S.V. Ley, J.C. Anderson, W.M. Blaney, P.S. Jones, Z.
    Lidert, E.D. Morgan, N.G. Robinson, D. Santafianos,
    M.J.S. Simmonds and P.L. Toogood, *Tetrahedron*, 1989,
    in press.
9. S.V. Ley, J.C. Anderson, W.M. Blaney, Z. Lidert,
    E.D. Morgan, N.G. Robinson, M.S.J. Simmonds,
    Tetrahedron Lett., 1988, *42*, 5433.
10. J.N. Nilton, P.S. Jones, S.V. Ley, N.G. Robinson and
    R.N. Sheppard, *Tetrahedron Lett.*, 1988, *29*, 1849.
11. S.V. Ley, A. Abad Somovilla, H.B. Broughton, D.
    Craig, A.M.Z. Slawin, P.L. Toogood and D.J.
    Williams, *Tetrahedron*, 1989, *45*, 2143.

New Strategies in the Synthesis of Pheromones and Antibiotics of the Avermectin-Milbemycin Groups

Y. Langlois
INSTITUT DE CHIMIE DES SUBSTANCES NATURELLES, C.N.R.S. 91190 GIF-SUR-YVETTE, FRANCE

The sexual pheromones of lepidoptera are generally constituted by a mixture, in accurate ratio, of several Z or E straight chain olefins or dienes bearing in ω position a functional group like an alcohol or an acetate.

Prior to about 10 years ago, classical methods for the formation of olefins, such as the Wittig reaction or the selective reduction of an acetylenic group, were generally employed for the synthesis of these pheromones.[1] However, in order to increase the geometrical purity of the target molecules, several other methods have now been introduced. For example the carbocupration of acetylenic derivatives[2] and the coupling of vinylic halides in the presence of transition metals[3] are now considered as highly selective and versatile methods.

The use of cyclic compounds as starting materials constitutes another possibility of efficiently controlling the geometry of a double bond. This methodology has been used, for example, in two syntheses of the juvenile hormone.[4] In this case, the geometry of a trisubstituted double bond was controlled using a dihydrothiopyran unit (1) (Scheme 1).

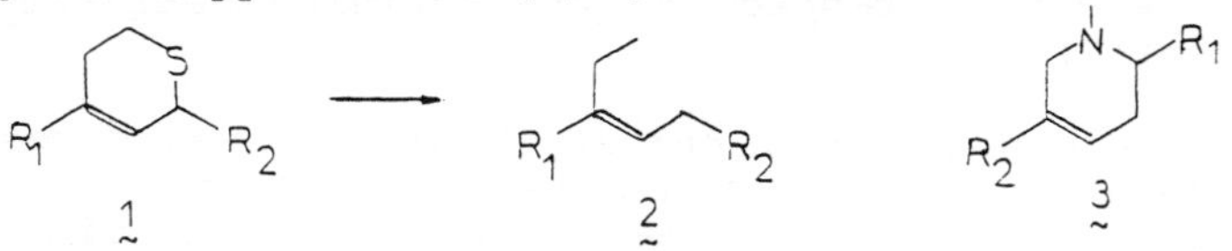

Scheme 1

Several years ago, when we commenced our research program towards the synthesis of various dienic pheromones, the use of such a methodology, but extended to tetrahydropyridinium salts, seemed to be very promising for several reasons. First, the tetrahydropyridines (3) are easily obtained in high yield from the corresponding pyridines which, in turn, are either commercially available or prepared just in a few steps. Secondly, we anticipated that an Hofmann elimination of a tetrahydropyridinium salt (7), in which the carbon 2 is substituted with a pseudo equatorial alkyl group, should give rise to a Z,E-dienamine by elimination of the proton antiperiplanar to the C-N bond (Scheme 2).

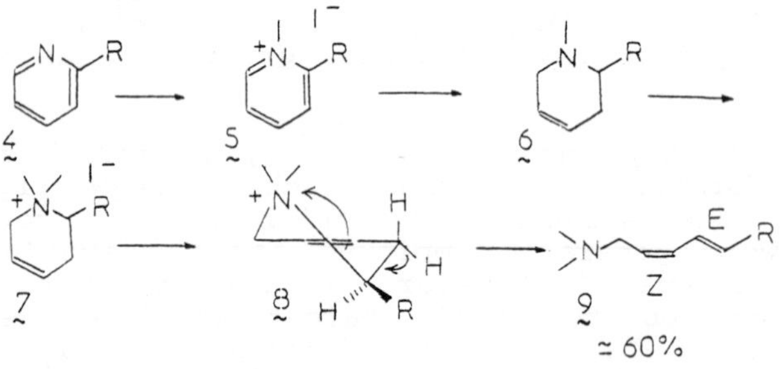

Scheme 2

This hypothesis was fully confirmed in the above sequence. Initial N-alkylation of 2-alkyl pyridine (4) followed by a classical $NaBH_4$ reduction of the pyridinium salt (5) to give a 2-alkyl-1,2,3,6-tetrahydropyridine (6) with good regioselectivity; quaternarisation of (6) to give the corresponding tetrahydropyridinium salt (7) followed by an Hofmann elimination affording a Z,E-dienamine (9) in 60% overall yield with respect to the pyridine derivative (4) (Scheme 2).

In a second stage of our synthesis, it was necessary to perform a regioselective alkylation of the dienamine (9) without isomerisation of the dienic unit. Initially we decided to use the method described by Schlosser.[5] Thus the dienamine (9) was treated sequentially with methyl chloroformate and tetrabutyl ammonium acetate to afford the Z,E-allylic acetate (11); no double bond isomerisation was evident. Acetate (11) was then alkylated with a Grignard reagent activated by lithium tetrachlorocuprate[6] to give rise to a mixture of Z,E and E,E dienes (12a) and (12b) in the ratio 90:10 and with 70-85% overall yield[7,8] (Scheme 3).

More directly, the Z,E-dienamine was alkylated with an excess of methyl iodide and the resulting ammonium salt (13) successfully alkylated with a Grignard reagent in the presence of the same copper salt.

Here again, we obtained a partial isomerisation of the Z double bond (Z,E/E,E = 90:10). It is worthy of note that no γ-substituted product was obtained in these reactions[8] (Scheme 3).

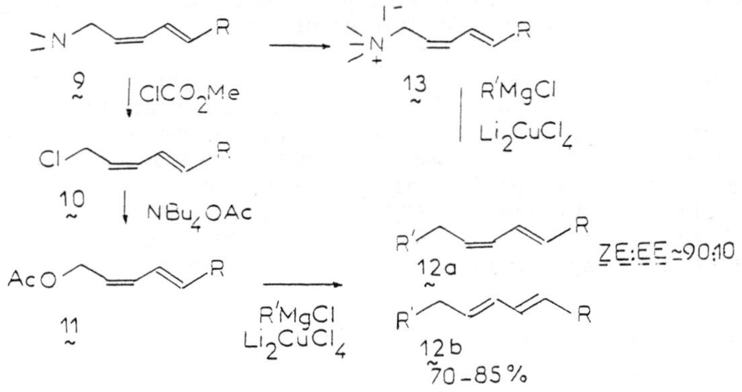

Scheme 3

The total synthetic sequence presented in the previous schemes can be illustrated by the preparations[8] of the 9Z,11E-tetradecadienyl acetate (18), one of the pheromones of Spodoptera littoralis and Spodoptera litura (Scheme 4).

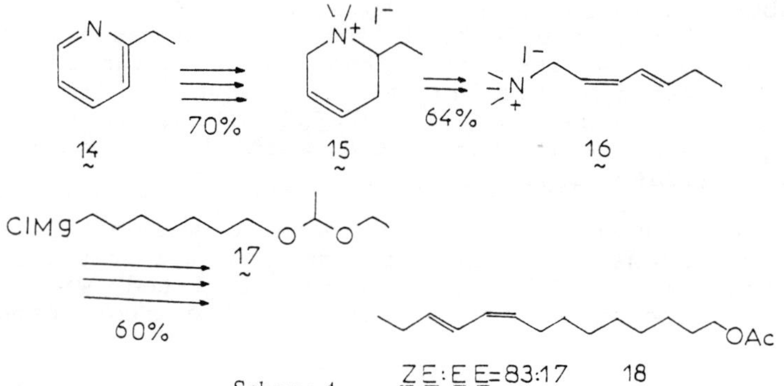

Scheme 4 $Z\,E\!:\!E\,E=83\!:\!17$ 18

A variable of this synthetic method is illustrated by the synthesis of the 9E,11E-dodecadienol
(27), the pheromone of the codling moth, Laspeyresia
pomonella (Scheme 5). The dimethylamino hexa-2Z,4E-
diene (9) (R = Me) was cleanly isomerised into the E,E
isomer (22) by the following sequence :

Oxidation of (9) with MCPBA and treatment of the
corresponding N-oxide (19) with trifluoroacetic anhydride to give a conjugated diene imminium (21) in the
more stable, all E, geometric configuration, and
regioselective reduction of the imminium with NaBH$_4$ to
yield the E,E-dienamine (22). Compound (22) was then
transformed into the pheromone (27) by the respective
sequence of reactions as described above.[8]

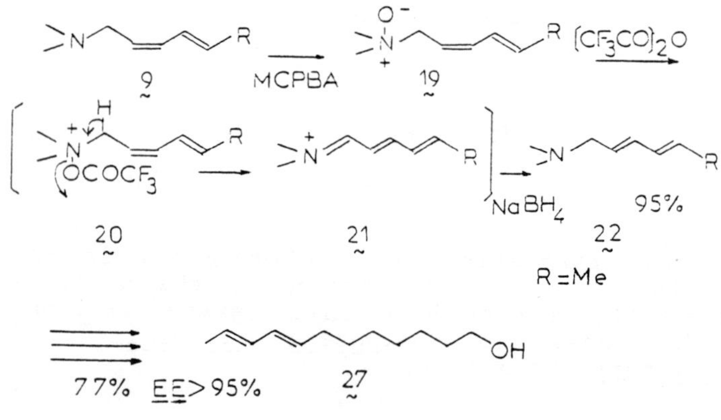

Scheme 5

$\underline{E},\underline{Z}$-dienic pheromones can also be synthesised by this versatile method. This is illustrated in the synthesis of the pheromone of Lobesia botrana, (25) a lepidoptera parasite of vineyards. In this sequence, pyridine (23), substituted at carbon 2 by a functionalised chain, was transformed into the corresponding dienic ammonium salt (24). This ammonium derivative was then alkylated either by an activated Grignard reagent as previously described or with lithium dialkyl organo cuprate. The yields and the selectivity of these alkylations were within the same range for both organometallic species[9] (Scheme 6).

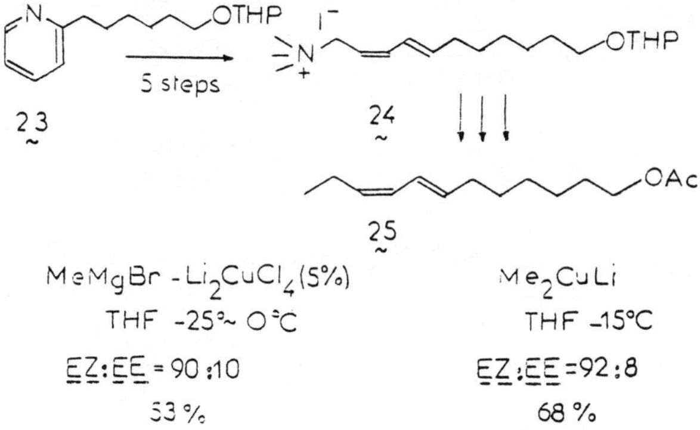

Scheme 6

The various syntheses achieved with this methodology are summarised in Scheme 7.

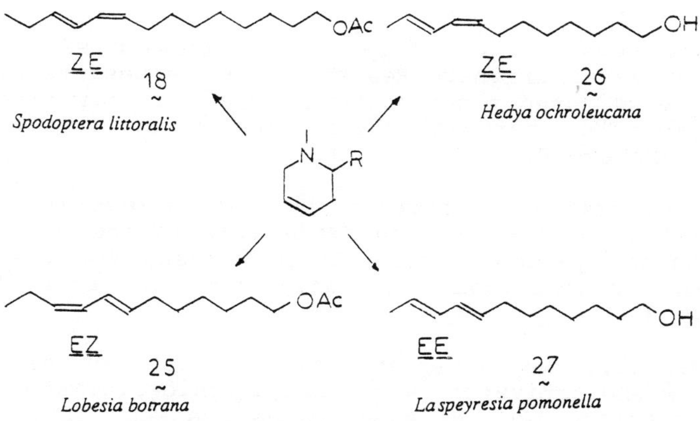

Scheme 7

In order to increase the selectivity of the alkylation step, we decided to undertake a systematic study
of the alkylation reaction of $\underline{Z}$-allylic derivatives.
Thus we tested the selectivity of the reaction with
various leaving groups and a variety of organometallic
species. To make a long story short, these studies
are summarised on the following scheme (Scheme 8).

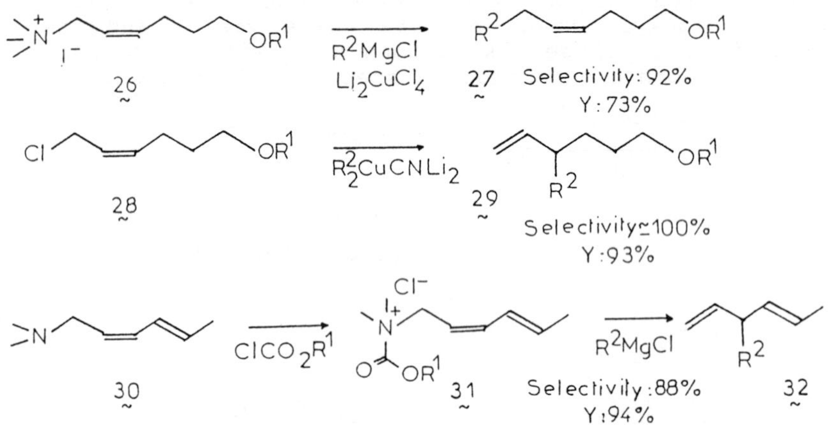

Scheme 8

The substitution of $\underline{Z}$-allylic ammonium salt (26)
with activated Grignard Reagent gave the best results;
the α-substituted product was nearly the only product
isolated. The stereoselectivity was about ninety per
cent[10] (Scheme 8).

A drammatic change in the regioselectivity was
observed with $\underline{Z}$-allylic chloride (28) in the presence
of lithium cyanocuprates. In this case, the γ-subs-
tituted product was the only product isolated[10]
(Scheme 8).

Finally, we observed that a tertiary allylic
amine when activated with alkyl chloroformate can
readily afford the SN'-substituted species (32). The

acyl ammonium salt intermediate (31) is much more reactive that the corresponding ammonium salt : the reaction was completed within one hour at -30°C compared to 24 hours at the same temperature with ammonium salts. It is also noteworthy that the regioselectivity was completely reversed[11] (Scheme 8).

A number of pheromones are constituted by a 1,4-dienic unit. Some examples are presented in Scheme 9.

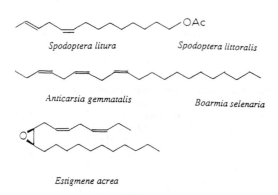

Spodoptera litura   *Spodoptera littoralis*

Anticarsia gemmatalis   *Boarmia selenaria*

Estigmene acrea

Scheme 9

For these species, we have again been interested to develop new synthetic methods, once more using the tetrahydropyridinium salts as starting material. It quite rapidly became evident that a silicon induced fragmentation could modify the regio selectivity of the Hofmann elimination.

It is well known that the Peterson olefination affords either Z or E olefins from β-hydroxy silanes according to the reaction conditions. We anticipated that the same type of reaction could be performed with tetrahydropyridinium salt bearing a silyl unit on the β-carbon of the side chain (Scheme 10).

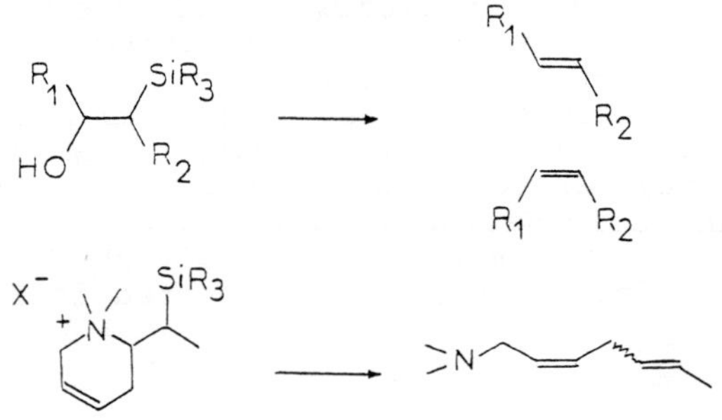

Scheme 10

To control the stereoselectivity of such elimination reactions it was necessary to obtain only one
diastereomer during the formation of the tetrahydropyridine (36) (Scheme 11). If we look at the mechanism of the reduction of the pyridinium salt (34), it
would appear that an hydride ion attacks stereo-
selectively the imminium salt intermediate (35). This
mechanism is more obvious if we consider the Newman
projection of this intermediate along $C_{1'}$-C_2 bond;
attack should give rise to the tetrahydropyridine (36)
of $2R^*$ $1'S^*$ relative configurations (Scheme 11).

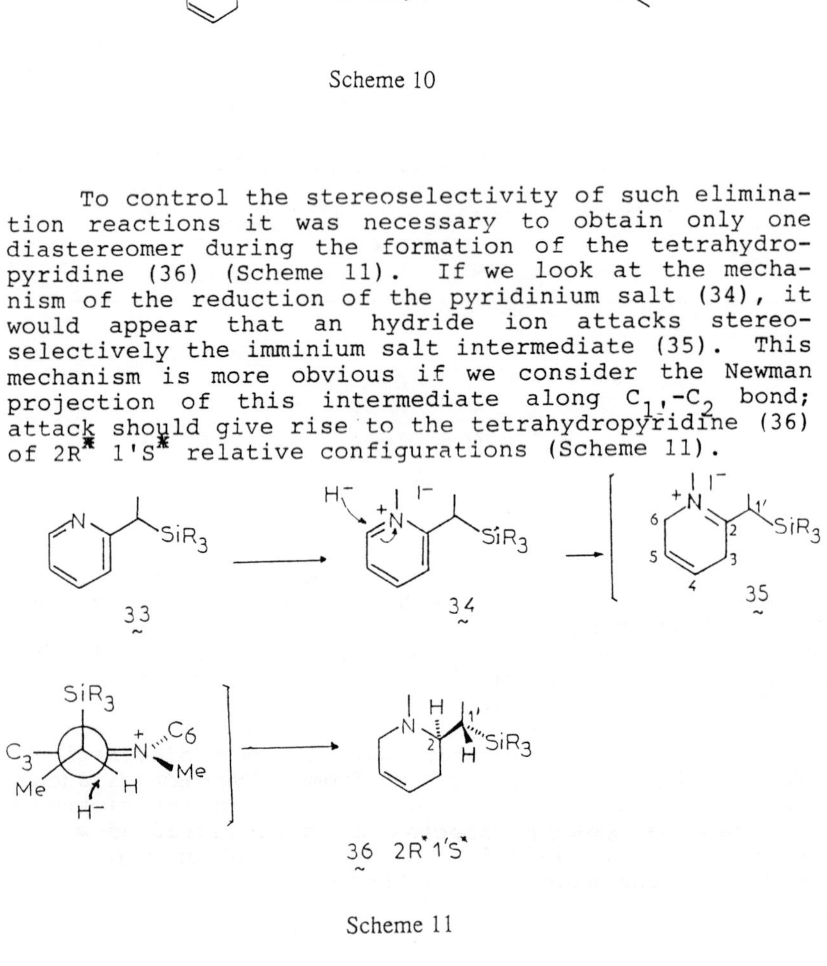

Scheme 11

Furthermore, this tetrahydropyridine (36) could then be subjected to two possible kinds of silicon induced fragmentations.

The first fragmentation method performed on the tetrahydropyridinium salt (37) (Scheme 12) and called "Sila-Hofmann" elimination is carried out in the presence of fluoride anion and is an anti-elimination giving rise to the 1,4-$\underline{Z}$,$\underline{E}$-dienamine (38).

The second method, which most probably arises via the $\underline{N}$-oxide intermediate (39), so called the "Sila-Cope" elimination, is a $\underline{syn}$-elimination affording the 1,4-$\underline{Z}$,$\underline{Z}$-dienamine (40) (Scheme 12).

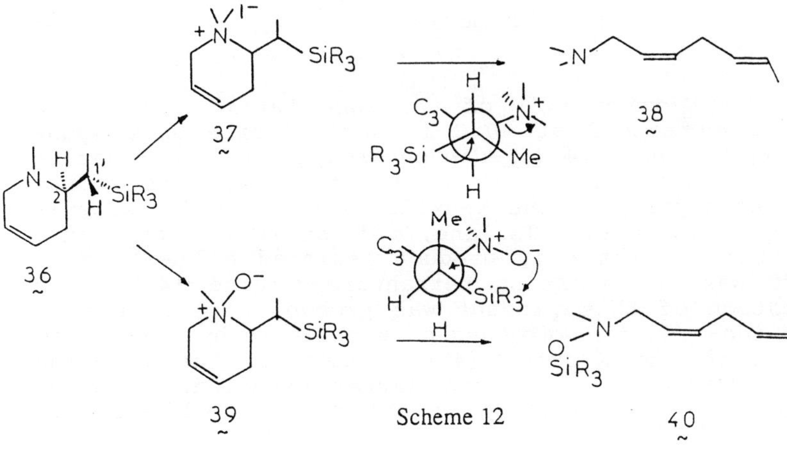

Scheme 12

Our hypotheses were fully confirmed.[12] The side chain of the 2-ethylpyridine ('14) was silylated and transformed, as previously described, into the corresponding tetrahydropyridinium salt (42). This compound, in the presence of cesium fluoride, gave rise, smoothly, to the anticipated 1,4-$\underline{Z}$,$\underline{E}$-dienamine (38). This dienamine (38), after the normal sequence of reactions, afforded the 9$\underline{Z}$,11$\underline{E}$-dodecadienyl acetate (43), pheromone of several species of lepidoptera (Scheme 13).

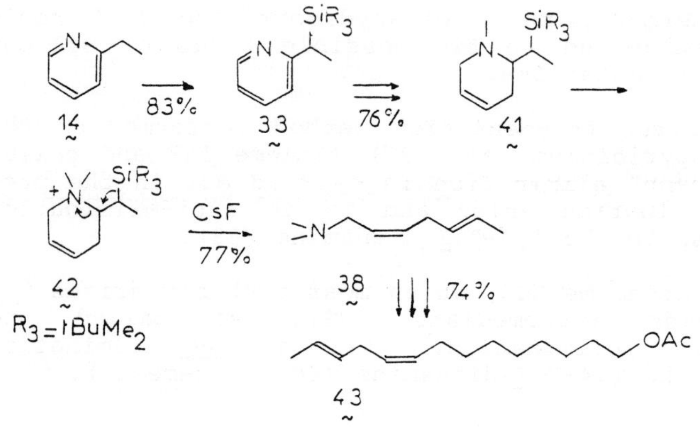

Scheme 13

The N-oxide (44) of the same tetrahydropyridine (41) when reacted at 40°C in acetonitrile gave spontaneously a 1,4-Z,Z-dienehydroxylamine derivative (45)[12] (Scheme 14). It is interesting to note that the tetrahydropyridine N-oxide (45), when not substituted with a trialkylsilane, did not afford any dienic derivative. The only product isolated by heating at 120°C was the starting tetrahydropyridine (46). The formation of this product was probably the result of an oxido-reduction process of the dienic hydroxylamine (48) with the N-oxide (47) present in the reaction medium (Scheme 14). This latter experience demonstrated the real utility of silicon in such eliminations.[12]

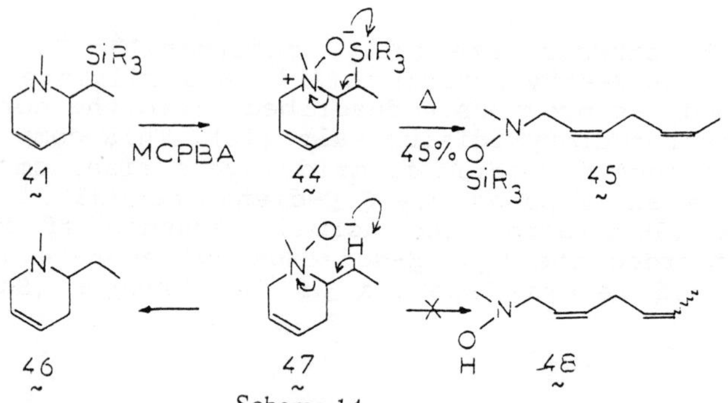

Scheme 14

This method of synthesis of 1,4 dienes was subsequently extended to the preparation of 1,5-dienes
namely a $_{10}$ contact pheromone of Drosophila
melanogaster^{10} (Scheme 14). The tetrahydropyridine
(50) was submitted to a Sila-Hofmann elimination. The
resulting 1,4-dienamine (51) was hydroxylated at the
terminal double bond and quaternarised. Alkylation
with a Grignard reagent was followed by oxidation of
the primary alcohol to an aldehyde. A Wittig reaction
with an appropriate phosphorane using the conditions
described by Bestmann^{13} afforded the anticipated
pheromone (54), identical in all respect with the
natural product (Scheme 15).

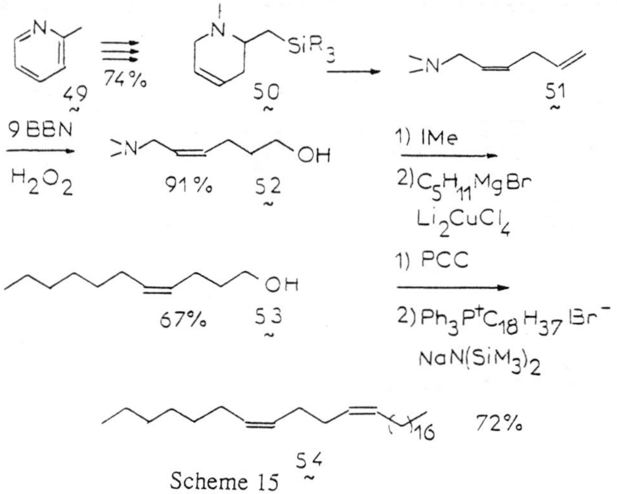

Scheme 15

An other example of this methodology was illustrated by the synthesis of the pheromone of
Pseudococcus comstockii.^{14} The retrosynthetic scheme
(Scheme 16) involved a [2,3] sigmatropic rearrangement
and a Sila-Cope elimination.

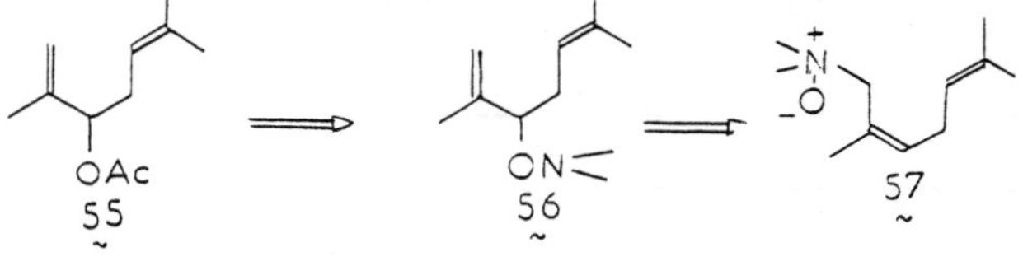

Scheme 16

Thus, 2,5-dimethyl pyridine (58) was sequentially alkylated and reduced in tetrahydropyridine (60). Oxidation of this compound with MCPBA gave rise to the corresponding $\underline{N}$-oxide (61) (Scheme 17) which, by heating in dichloromethane, afforded three compounds (Scheme 17). The expected diene (62), resulting from a normal Sila-Cope elimination and two other products (63) and (64) resulting of an unexpected 1,2 carbon-carbon silicon shift were observed. In acetonitrile only two products were formed, the diene (62) and the eight-membered ring compound (63). These quite surprising results are probably due to a peculiar conformational effect resulting from high steric hindrance around nitrogen and carbon 1' in $\underline{N}$-oxide (61).

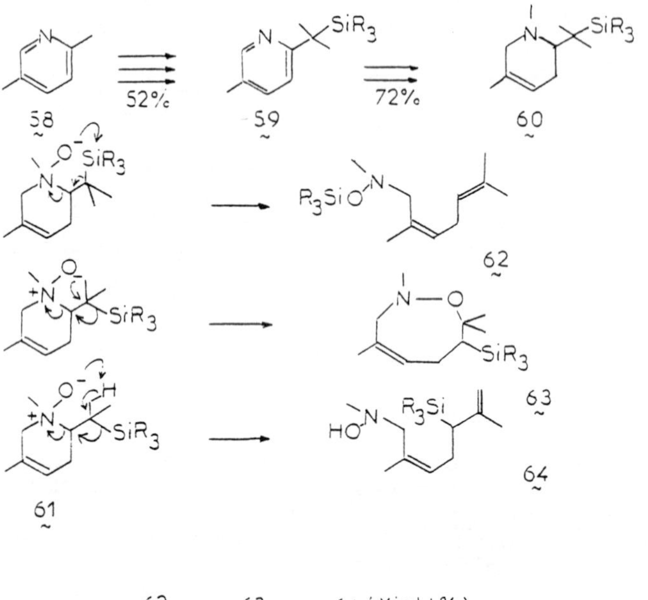

	62	63	64 (Yield %)
CH_2Cl_2	25	50	25
CH_3CN	50	50	0

Scheme 17

'Both compounds (62) and (63) were subsequently transformed into the dienic N-oxide (57) by sequential treatment by methyl iodide and cesium fluoride (Scheme 18). A [2,3] sigmatropic rearrangement gave rise to compound (56) which afforded the pheromone (55). The dienic N-oxide (57) has also been transformed into the corresponding dienic primary alcohol, an intermediate in the enantioselective synthesis of the pheromone (55) by Mori.[15]

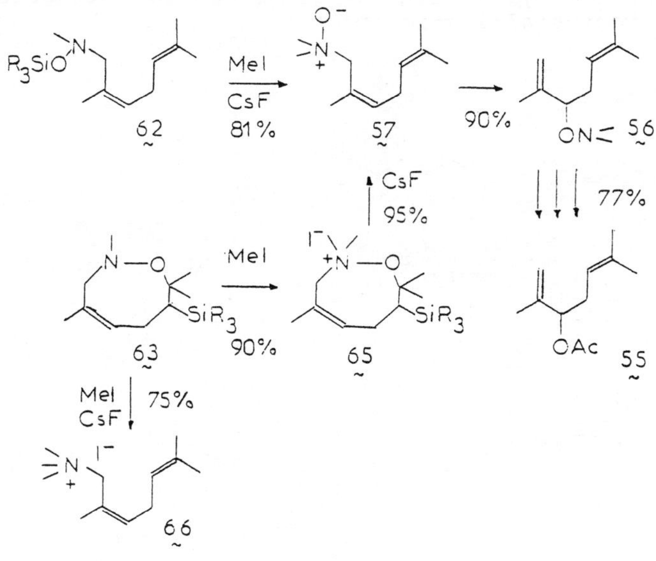

Scheme 18

In order to test the possibility of synthesising trienic Z,Z,Z 1,4,7 units, characteristic of several pheromones including (67), pheromone of Boarmia selenaria, we prepared the tetrahydropyridine (69) (Scheme 19). This compound was oxidized by MCPBA to afford two diastereomeric N-oxides (70) and (72). In this particular case, these two compounds were stable

enough to be isolated. We observed that one N̲-oxide
selectively gave rise to the product (71) resulting by
the anticipated Sila-Cope elimination, and the other
by the classical Cope elimination gave the dienic
compound (73). At that period, however, we did not
know the configuration at the nitrogen centre and the
two configurations at carbon 2 and carbon 1' were only
speculative-deduced from examination of molecular
models. So we decided to undertake an X̲-ray analysis
on one of these tetrahydropyridines N̲-oxides, (70) or
(72). Unfortunately, these compounds were not stable
sufficiently and gave rise to decomposition products
during the recrystallisation procedure.

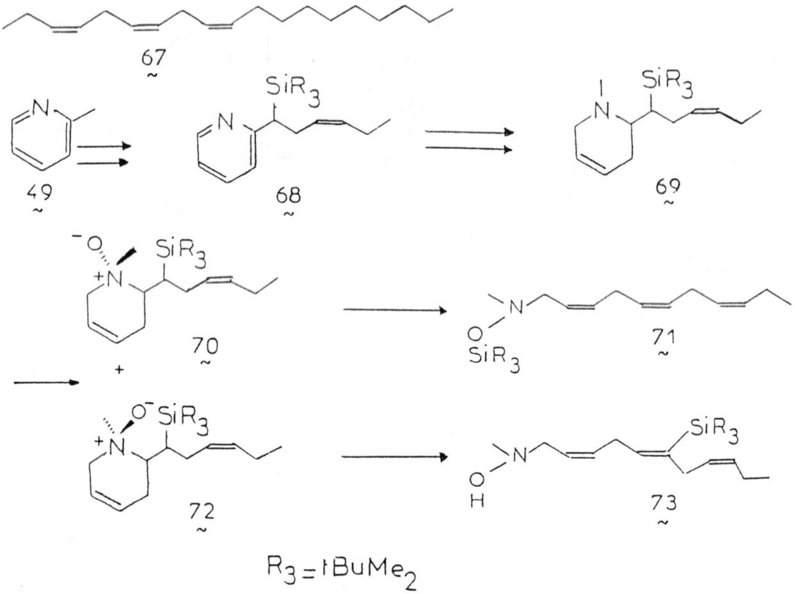

$R_3 = tBuMe_2$

Scheme 19

On the other hand, the tetrahydropyridinium salts
were more stable. Hence, we decided to prepare the
tetrahydropyridinium salt (75) bearing a cyano

methylene unit on nitrogen which could be deprotonated to give rise to a <u>syn</u>-elimination by a mechanism similar to the Sila-Cope elimination.

Thus, the tetrahydropyridine (74) was N-alkylated with iodo acetonitrile and afforded one diastereomeric tetrahydropyridinium salt (75) as the major compound. Treatment of this salt with potassium carbonate in dimethyl formamide gave rise to an elimination reaction involving a proton and not the silicon unit (Scheme 20).

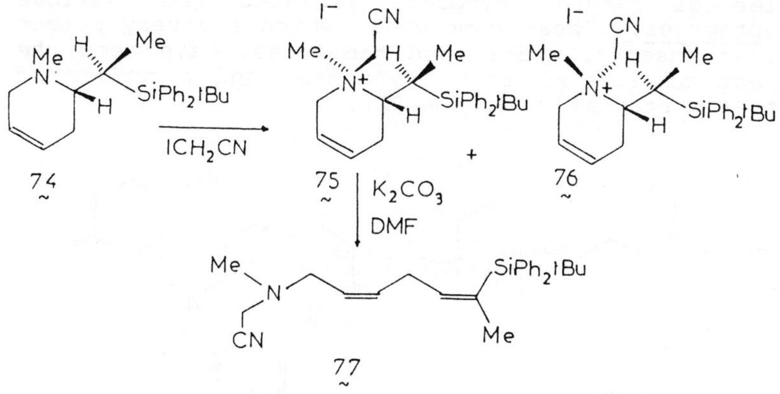

Scheme 20

X-ray analysis of a single crystal of the tetrahydropyridinium salt (75) not only gave the relative configuration at nitrogen, but also confirmed that the two configurations at carbon 2 and carbon 1' were as we had previously anticipated (Scheme 21).

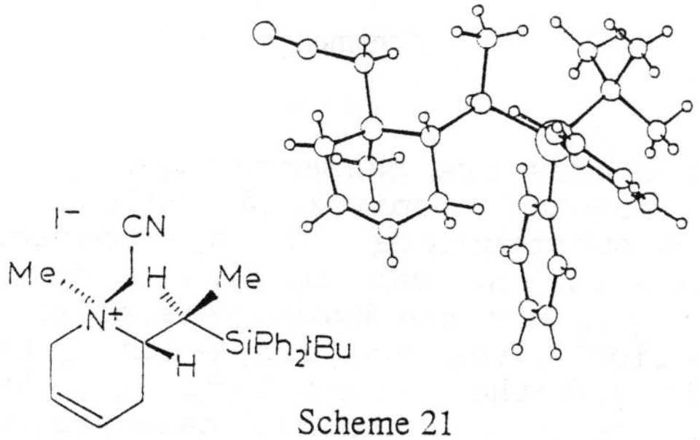

Scheme 21

From this result we can also deduce the configuration at nitrogen for the two diastereomeric N-oxides
(70) and (73).

Further synthetic studies concerning those sili-
cone induced fragmentations are now in progress in our
laboratory.

Avermectins (1)[16] and Milbemycins (2) are two
series of natural products isolated from various
Streptomyces. These compounds, which are very potent
against insects, acari and nematodes, have been the
subject of several total syntheses and a number of
synthetic studies[18] (Scheme 22).

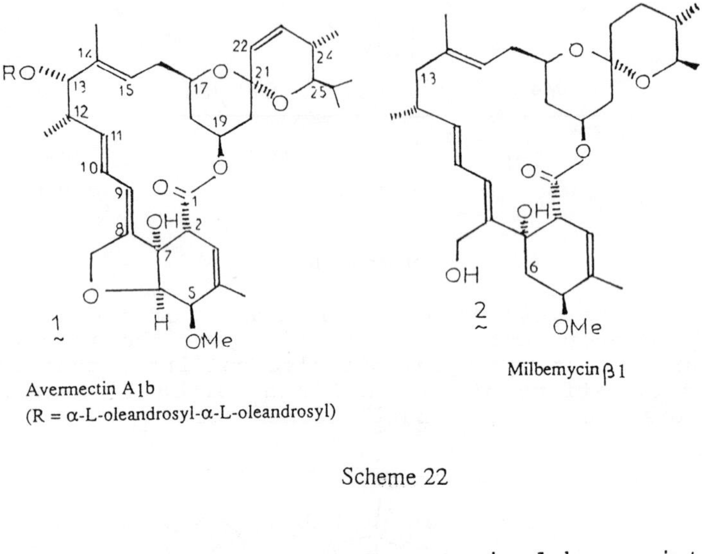

Avermectin A1b
(R = α-L-oleandrosyl-α-L-oleandrosyl)

Milbemycin β 1

Scheme 22

Their structures are characterised by a sixteen-
membered ring lactone, a spiroketal unit and an hexahydrobenzofuran substructure. In one possible retrosynthetic analysis, one can imagine a disconnection
between C_{14} and C_{15} and the hydrolysis of the lactone.
This disconnection gives two fragments : the spiroketal unit (5) and the larger part (6), the latter
retaining both the dienic and the hexahydrobenzofuran

entities   (Scheme 23). In a following disconnection, the dienic, "western" part of the molecule (6)
could be the result of an Hofmann elimination from the
tetrahydropyridinium salt (8), whereas the hexahydrobenzofuran could be formed by a Diels-Alder reaction
or a Robinson type annelation.

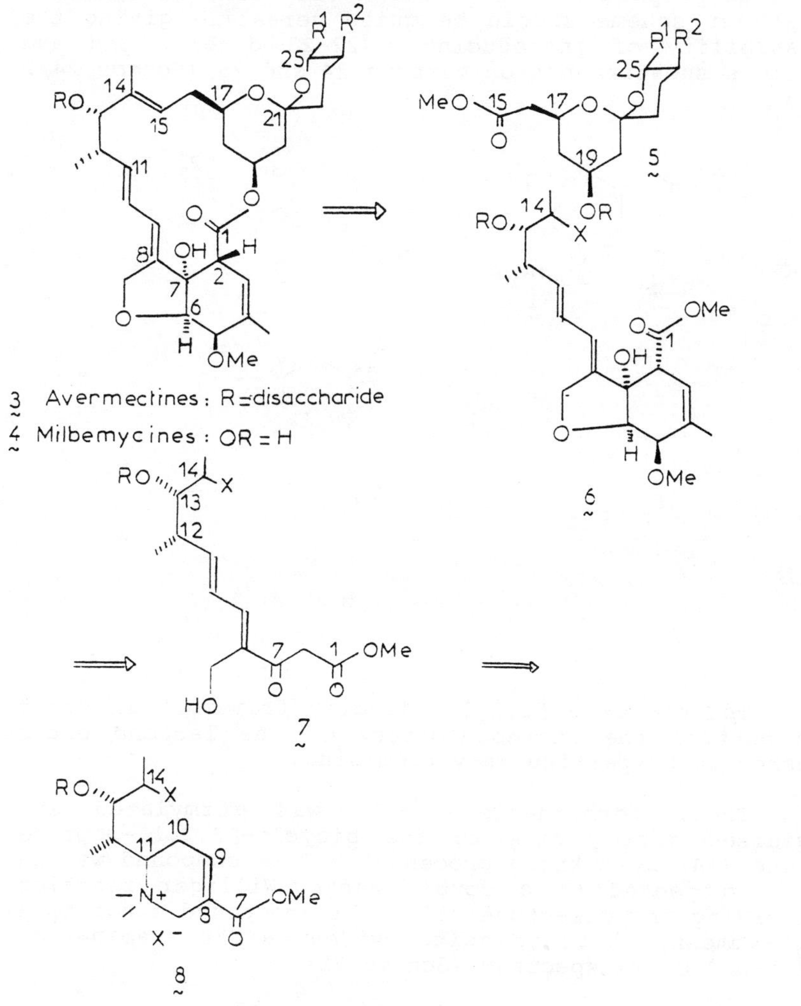

3 Avermectines : R = disaccharide

4 Milbemycines : OR = H

Scheme 23

Initially, we decided to study the synthesis of the spiroketal fragment of the molecule by using nucleophilic attack of lithium acetylide, formed by deprotonation of the homopropagylic alcohol (10), on the lactone ester (9). Among a number of possible ways to prepare this bicyclic ketal (11), we believed that our scheme should be quite versatile giving the possibility of introducing a 22-23 double bond and various substituents on carbons 24 and 25 (Scheme 24).

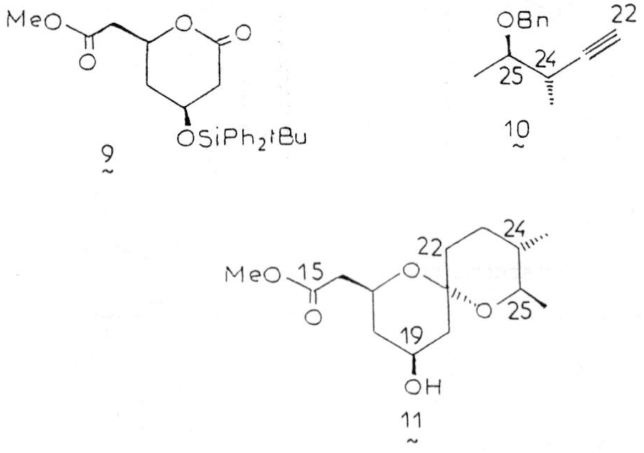

Scheme 24

The use of a [2,2,1] bicyclic framework in order to control the stereochemistry of the lactone ester seemed to be particularly promising.

Thus, norbornadiene (12) was formylated and oxidised giving rise to the bicyclo-[2,2,1]-heptane dione (14) by a known process.[19] This compound was in turn subjected to a double Baeyer-Williger reaction affording the dilactone (15). This compound exhibits a C_2 symmetry which is quite evident after examination of its ^{13}C NMR spectrum (Scheme 25).

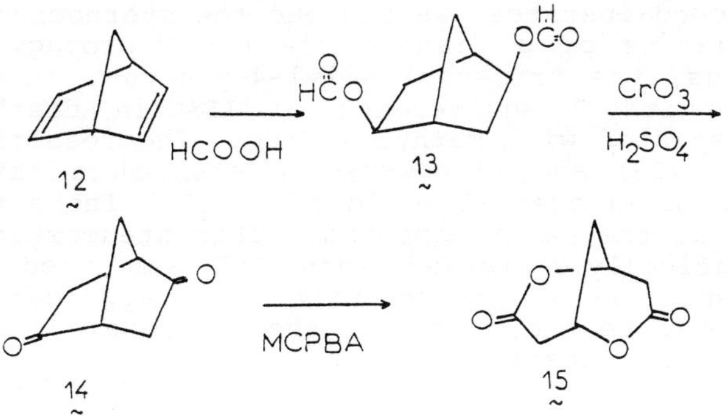

Scheme 25

Methanolysis of dilactone (15) afforded the diester diol (16) which by lactonisation led to the lactone ester (17) (Scheme 26).

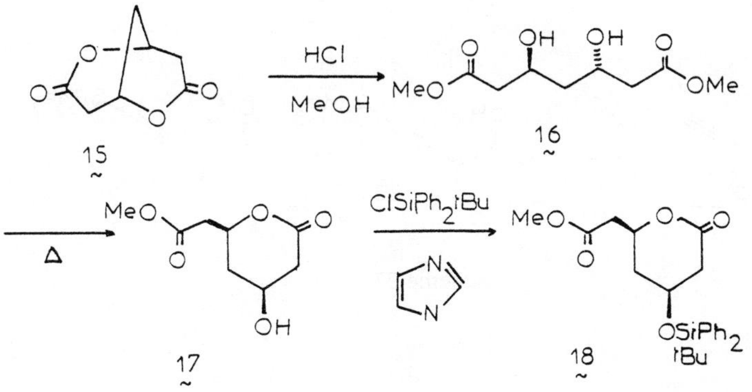

Scheme 26

In a second instance, we studied the stereochemistry of alkylation of a dianion of an homopropagylic alcohol. Thus, the trimethyl silyl-4-pentyn-2-ol was deprotonated with 2 equivalents of LDA in diethyl ether and alkylated with methyl iodide. The resulting diastereomers (23) and (24) were isolated in a ratio of 90:10 and in an overall yield of 70%.[20] The anti-isomer (23) was the major compound. This stereoselectivity dramatically decreased when THF was used as solvent. Hence we suspected that some aggregation phenomenon could be the origin of the stereoselectivity of this reaction (Scheme 27).

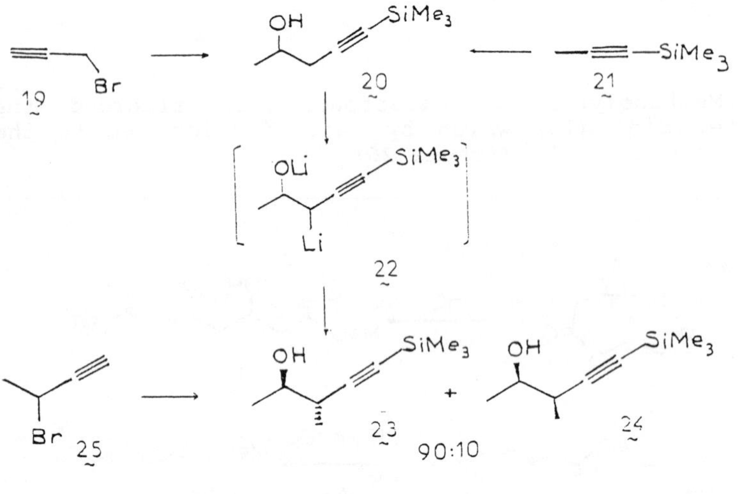

Scheme 27

There is a dual interest in this reaction. The primary lies in the possibility of preparing enantiomerically pure homopropagylic alcohols employing the method described by Yamamoto.[21] The second interest is the potential versatility of this method which allows introduction of a large variety of R groups on C_{25} (from an aldehyde) and R' on C_{24} (from an alkyl-halide).

After deprotection of the acetylenic group and protection of the alcohol, the lithium acetylide obtained from compound (10) was condensed with a good chemioselectivity on the lactone ester (18). The acetylenic ketone intermediate (26) after reduction of the triple bond and simultaneous deprotection of the alcohol afforded the target spiroketal (11)[20] (Scheme 28).

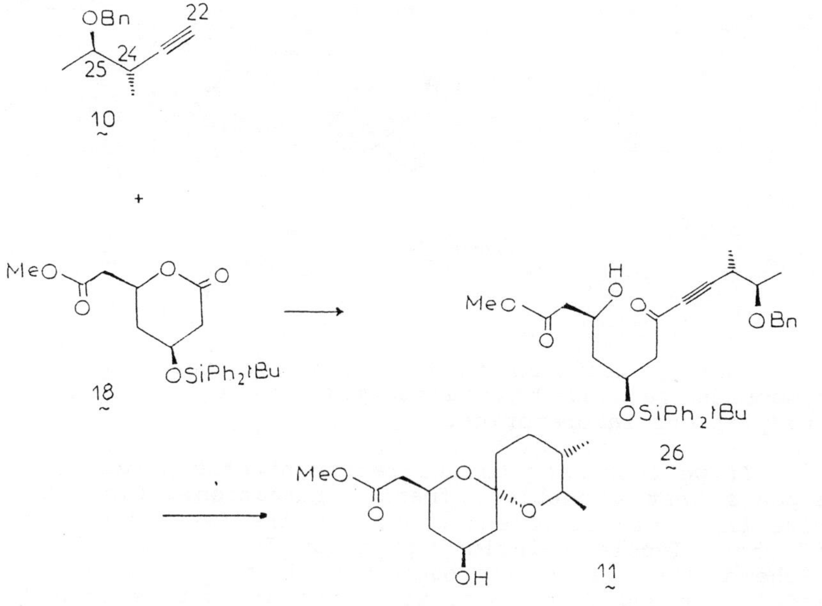

Scheme 28

Further studies are current in our laboratory in order to prepare the bicycloheptanedione (14) as an enantiomeric pure compound by virtue of an asymmetric Diels-Alder reaction.[22] The use of chiral α,β-unsaturated oxazoline, which after activation with trifluoro acetic anhydride is a very powerful dienophile, seems to be very promising route. This is illustrated in Scheme 29.

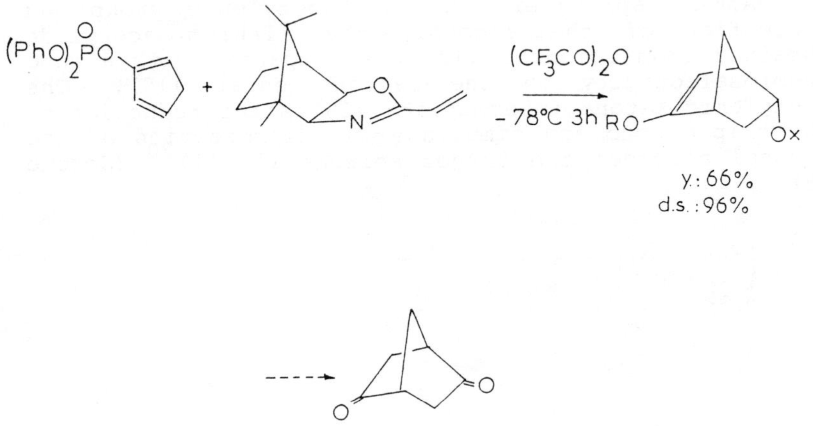

y: 66%
d.s. : 96%

Scheme 29

Another approach to the total synthesis of the compounds related to avermectins is equally under study in our laboratories.

If we look back to the retrosynthetic scheme, it appears that a double Hofmann elimination could give rise in a very efficient manner to the "western" part of the molecule - including carbon 7 to carbon 17 - (Scheme 30). This approach has been studied on a model compound free of substituents at carbons 12 and 13.

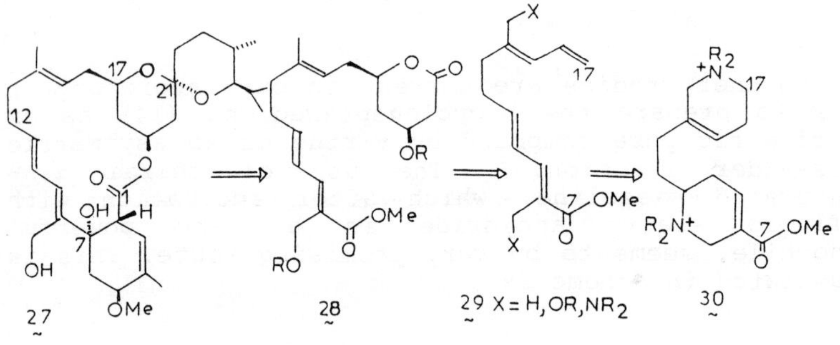

Scheme 30

Thus, 5-methyl methylnicotinate (32) was condensed with nicotinaldehyde (33) affording the crystalline compound (34). The latter was hydrogenated, N-alkylated and then reduced with NaBH$_4$ to afford a bis tetrahydropyridine (36) (overall yield 56%) (Scheme 31).

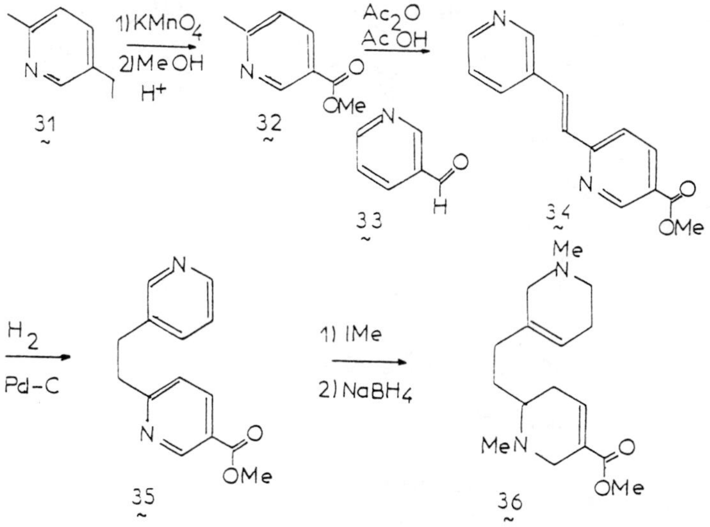

Scheme 31

After reduction of the ester group the corresponding alcohol (37) was N-alkylated and subjected to an Hofmann elimination. This reaction afforded the tetraenic compound (39) in which the framework from carbon 7 to carbon 17 of the avermectins was included with a perfect control of the geometry of three double bonds (Scheme 32).

Further studies are presently in development in order to introduce two substituents on carbons 12 and 13 and to control the chirality of these centres.

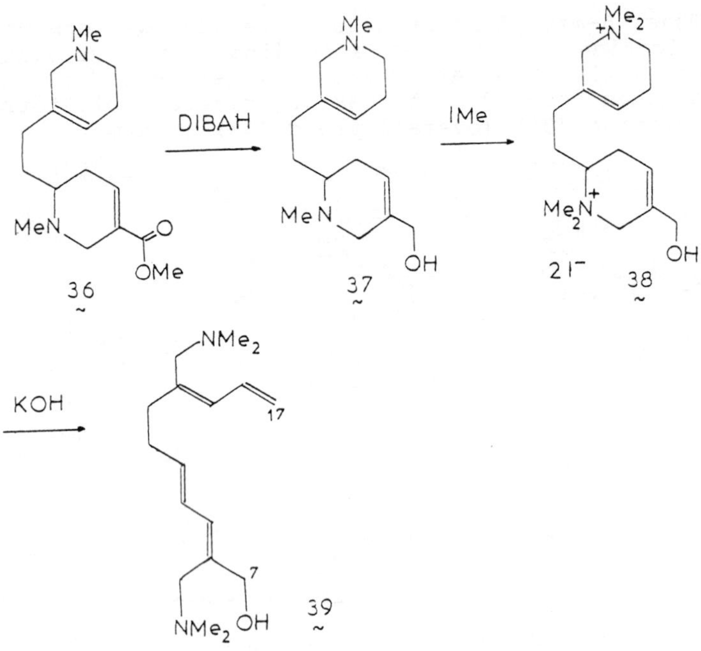

Scheme 3 2

In conclusion, this work has demonstrated the utility of the Hofmann elimination of tetrahydropyridinium salts in order to control the geometric configuration of polyolefinic compounds.

During the synthesis of the spiroketal unit of avermectins and milbemycins, a novel strategy involving the use of a [2,2,1] framework as well as a highly stereoselective alkylation of the dianion of homopargylic alcohol have been studied. The enantioselective synthesis of this unit <u>via</u> an asymmetric Diels-Alder reaction with a chiral oxazoline is now in process of development.

Last but not least, it is a great pleasure to thank my fellow colleagues who made all this work successful: Dr. N.V. Bac, Dr. Y. Fall, Dr. G. Dressaire and Dr. L. Konopski.

REFERENCES

1. a) R. Rossi, Synthesis, 1977, 817. b) C.A. Henrick, Tetrahedron, 1977, 33, 1845. c) For the synthesis of chiral pheromones see: K. Mori, Tetrahedron, 1989, 45, 3234.
2. A. Alexakis, J. Normant, Synthesis, 1981, 841.
3. R.F. Heck, 'Palladium Reagents in Organic Synthesis', Academic Press, New-York, 1985.
4. a) K. Kondo, A. Negishi, K. Matsui and D. Tunemoto, J.C.S. Chem. Comm., 1972, 1311. b). J.P. Demoute, D. Hainaut and E. Toromanoff, C. R. Acad. Sc., Serie C, 1973, 277, 49.
5. G. Fouquet and M. Schlosser, Angew. Chem. Int. Ed., 1974, 13, 82.
6. M. Tamura and J. Kochi, Synthesis, 1971, 303.
7. D. Samain, C. Descoins and Y. Langlois, Nouv. J. de Chimie, 1978, 2, 249.
8. C. Descodts, G. Dressaire and Y. Langlois, Synthesis, 1979, 510.
9. G. Dressaire and Y. Langlois, Tetrahedron Letters, 1980, 67.
10. N.V. Bac, Y. Fall and Y. Langlois, Tetrahedron Letters, 1986, 27, 841.
11. Y. Langlois, N.V. Bac and Y. Fall, Tetrahedron Letters, 1985, 26, 1009.
12. N.V. Bac and Y. Langlois, J. Am. Chem. Soc., 1982, 104, 7666.
13. H.J. Bestmann, W. Spransky and O. Vostrowsky, Chem. Ber., 1976, 109, 1694.
14. Y. Fall, N.V. Bac and Y. Langlois, Tetrahedron Letters, 1986, 27, 3611.
15. K. Mori and H. Ueda, Tetrahedron, 1981, 37, 2581.
16. G. Albers-Schönberg, B.H. Arison, J.C. Chabala, A.W. Douglas, P. Eskola, M.H. Fischer, A. Lusi, H. Mrozik, J.L. Smith and R.L. Tolman, J. Am. Chem. Soc., 1981, 103, 4216.
17. H. Mishima, M. Kurabayashi, C. Tamura, S. Sato, H. Kuwano and A. Saito, Tetrahedron Letters, 1975, 711.
18. a) H.G. Davies and R.H. Green, Nat. Prod. Rep., 1986, 3, 88. b) G. Quinkert, Synform, 1986, 4, 1. c). S.J. Danishefsky, D.M. Armistead, F.E. Wincott, H.G. Selnick and R. Hungate, J. Am. Chem. Soc., 1989, 111, 2967 and references therein.
19. R.T. Hawkins, R.S. Hsu and S.G. Wood, J. Org. Chem., 1978, 43, 4648.

20. N.V. Bac and Y. Langlois, Tetrahedron Letters, 1988, 29, 2819.
21. N. Ikeda, I. Arai and H. Yamamoto, J. Am. Chem. Soc., 1986, 108, 483.
22. A. Pouilhès, E. Uriarte, C. Kouklovsky, N. Langlois, Y. Langlois, A. Chiaroni and C. Riche, Tetrahedron Letters, 1989, 30, 1395.

Synthesis of Some Fluorinated Non-Ester Pyrethroids

M. J. Bushell

IMPERIAL CHEMICAL INDUSTRIES, CHEMISTRY DEPARTMENT, JEALOTT'S HILL
RESEARCH STATION, BRACKNELL, BERKS. RG12 6EY, UK

1. INTRODUCTION

The outstanding contribution to Agrochemical research
made by the Rothamsted group led by Michael Elliott is
well known[1].

Their discoveries have led to the development of
several excellent insecticides including permethrin,
cypermethrin and deltamethrin. In addition their
success helped initiate a major research effort
throughout the agrochemical industry searching for new
products providing further technical advantages.

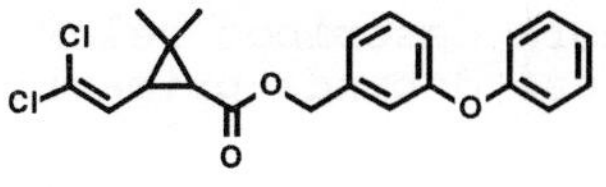

Permethrin

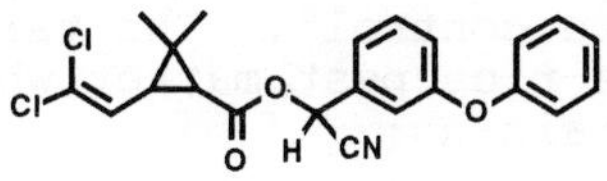

Cypermethrin

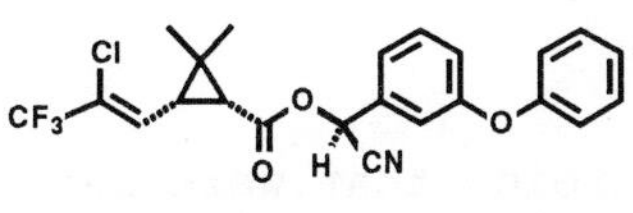

Lambdacyhalothrin

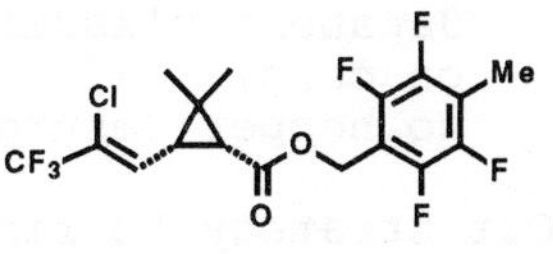

Tefluthrin

Figure 1. Pyrethroids commercialised by ICI.

ICI have been engaged in pyrethroid research for
many years. We have been involved in the
commercialisation of permethrin and cypermethrin. Our
in-house efforts were rewarded first by the discovery
of cyhalothrin[2], launched initially as an animal health
product (Trademark Grenade), then as the enantiomer
pair lambda-cyhalothrin (Trademark Karate) for
agricultural use[3]. This compound is more active than
cypermethrin and has important additional benefits (e.g
prevention of spider mite resurgence owing to much
improved acaricidal activity).

The most important markets not satisfied by the
Rothamstead NRDC pyrethroids are:

(i)   Soil pests   -   due to poor bioavailability
(ii)   Rice pests   -   due mainly to very high
                            toxicity to fish in
                            laboratory tests.
(iii)   Spider mites -   due to low activity.

Many companies have been attempting to extend the
utility of pyrethroids to these new crop/pest
applications, with some notable successes. ICI have
developed tefluthrin (PP993, Trademark Force) as the
first soil active pyrethroid[4] and FMC have launched
biphenthrin (FMC 54800, Trademark Talstar) for spider
mite control[5]. The target of work described below is
the rice pest market where success depends primarily on
obtaining:

(i)   High activity on rice pests (especially plant
         and leaf hoppers, rice water weevil,
         stemborers).
(ii)   Fish safety (preferably category A in the
         Japanese classification system, TLm (48h)>10ppm
         on carp).
(iii)   No hopper resurgence.

Our strategy to find pyrethroids that were safer
to fish initially concentrated on investigation of a
wide variety of acid and alcohol combinations.
Significant reductions in toxicity to fish were
discovered (by ourselves and others[6-8]) with 4'-halo or
α-ethynyl substitutions in the alcohol moiety, or by
"addition" of halogens (or HX) across the vinyl group
on the cyclopropane ring.

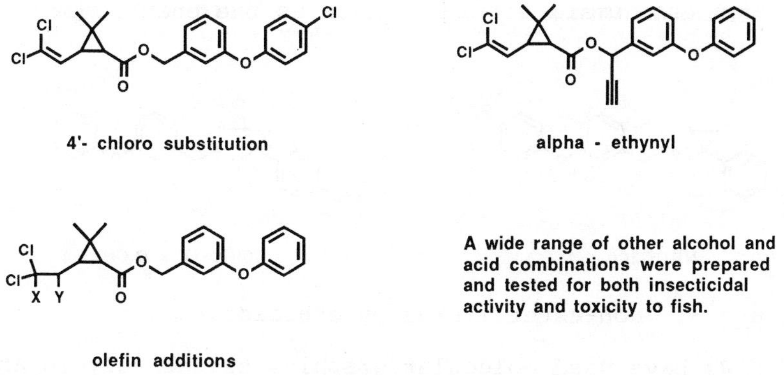

4'- chloro substitution

alpha - ethynyl

olefin additions

A wide range of other alcohol and
acid combinations were prepared
and tested for both insecticidal
activity and toxicity to fish.

Figure 2. Pyrethroids with reduced toxicity to fish.

Despite these encouraging leads, it eventually
proved impossible to combine high levels of
insecticidal activity on plant hoppers with sufficient
levels of safety to fish to meet the target
requirements by this strategy.

Mitsui Toatsu[9] made an important breakthrough with
the discovery of non-ester pyrethroids MTI-500
(commercialised as Etofenprox) and MTI-800. These
compounds have served as an important lead area with
many companies actively involved. NRDC 200[10] and the
difluorocyclopropyl analogues from CSIRO[11] are examples
of analogues with increased activity.

In describing these non-ester linked pyrethroids
reference will be made to "acid" and "alcohol" portions
of the molecule. This is to facilitate comparisons
with traditional ester-linked pyrethroids and
(hopefully) should not be a source of confusion. The
ethoxyphenyl ring of MTI-500 and MTI-800 is clearly the
"acid end".

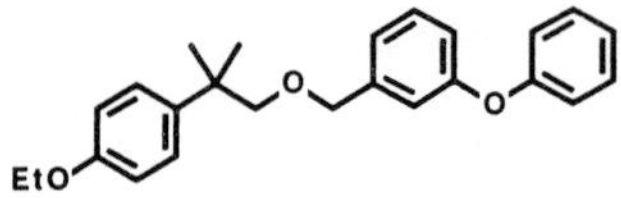

Etofenprox (MTI500)

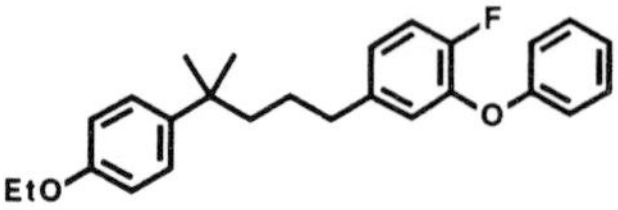

MTI800

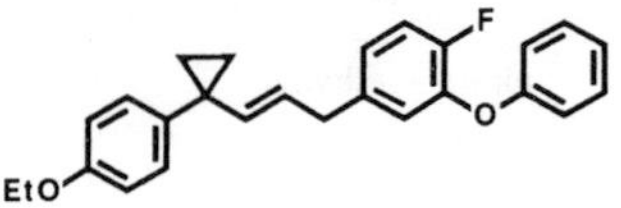

NRDC200

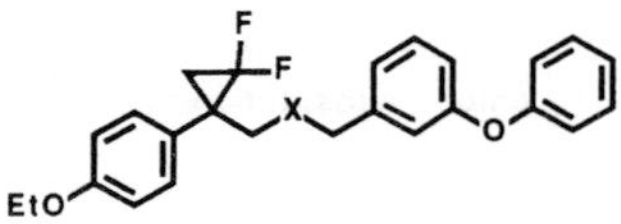

CSIRO (X = O,CH$_2$,S)

Figure 3. Non-ester linked pyrethroids.

We have used molecular graphics extensively in an attempt to understand the key features of the pyrethroid toxophore. A "template" has been constructed[12] to rationalise the activity of major pyrethroid types. Satisfactory overlays of Mitsui-types can be achieved provided only one methyl group is required to closely match the (pro-S)-methyl group in the cyclopropane ring of permethrin (known to be important for activity).

The modelling produced a number of ideas that led to the synthesis of a number of active series[13] - and some inactive ones! For the work described below our belief that only one methyl group is important for activity is worthy of note.

Fluorine Substitution and Biological Activity

Fluorine for hydrogen substitution is a tactic often employed when searching for improvements in biological activity. Such substitutions can cause large potency increases, but sites where fluorine will increase activity cannot be predicted with certainty.

The effect of fluorine incorporation on biological activity is perhaps best encapsulated by Colin Swithenbank[14] (Rohm and Haas). He has reported many examples which show that "replacement of H by F usually lowers activity BUT the most active compounds often contain fluorine somewhere".

Thus to be sure of preparing the most active compounds in a series it may be necessary to selectively fluorinate the molecule everywhere.

Two additional general observations can be made from personal experience:

(i)   Large perfluoroalkyl groups often lower activity.
(ii)   Fully halogenated alkyl groups are usually best e.g. CF_3 usually better than CH_2F, CHF_2.

Inspection of the structure of MTI-800 for example, reveals that there are a number of sites to investigate the effect of fluorine substitution:

(1) The linking chain
(2) The "alcohol" aromatic rings
(3) The "acid" aromatic ring
(4) The ethoxy substituent
(5) The gem-dimethyl group

We have made F for H substitutions in all areas of the molecule (although not all possibilities have been prepared). Generally activity has been lowered, but large increases in insecticidal activity have been found in some cases, with substitutions in the gem-dimethyl group of particular interest.

A summary of the results we have obtained is shown in Figure 4.

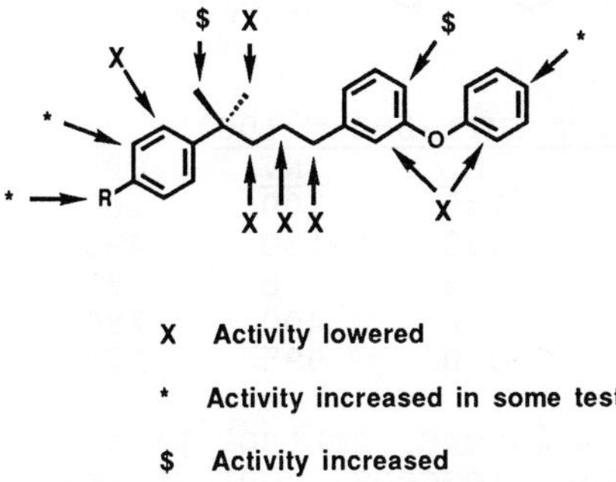

X   Activity lowered

*   Activity increased in some tests

$ Activity increased

Figure 4.   Summary of the effect of fluorine on insecticidal activity.

(1) F for H in the linking chain lowered activity in
     all examples prepared.

(2) In the "alcohol" moiety no deviations from
     traditional pyrethroid SAR were apparent, with 4-F
     substitution (as in the "Bayer alcohol") best.

(3) Fluorines <u>ortho</u> to the ethoxy group increased
     activity on some pests - particularly mites and
     lepidoptera, but lowered activity on brown
     planthopper. 4-Substituted 2,3,5,6-tetrafluoro
     analogues were inactive.

(4) Fluorinated <u>para</u> substituents <u>e.g</u>. $4-OCF_3$
     increased activity in some cases. $4-OCF_2Cl$ is
     known to enhance mite activity in the MTI-500
     series.

(5) The rest of the presentation will describe various
     replacements for the gem-dimethyl group, and the
     effect on insecticidal activity.

The insecticidal activity of two early lead
compounds[15-17] (1a, 1b) is shown in Figure 5. These
have a single CF_3 group replacing the gem-dimethyl in
MTI-500 and differ only in the alcohol 4-fluorine
substitution. As can be imagined this activity proved
very exciting. For comparison the monomethyl
analogue of MTI-500 would have a potency ratio of 10-40
in these tests. Thus CF_3 replacement for methyl has
led to a greater than 10-fold increase in potency.

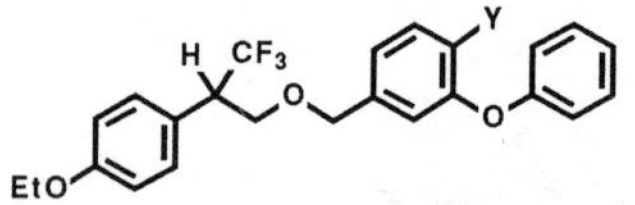

Relative Potency (MTI-500 = 100)

	NL	NV	CP
MTI-500	100	100	100
MTI-800	170	150	490
Cycloprothrin	2	8	30
(1a)	380	400	280
(1b)	370	995	220

NL <u>Nilaparvata lugens</u>, contact test
NV <u>Nephotettix virescens</u>, contact test
CP <u>Chilo Partellus</u>, residual test

Figure 5. Insecticidal activity of lead compounds

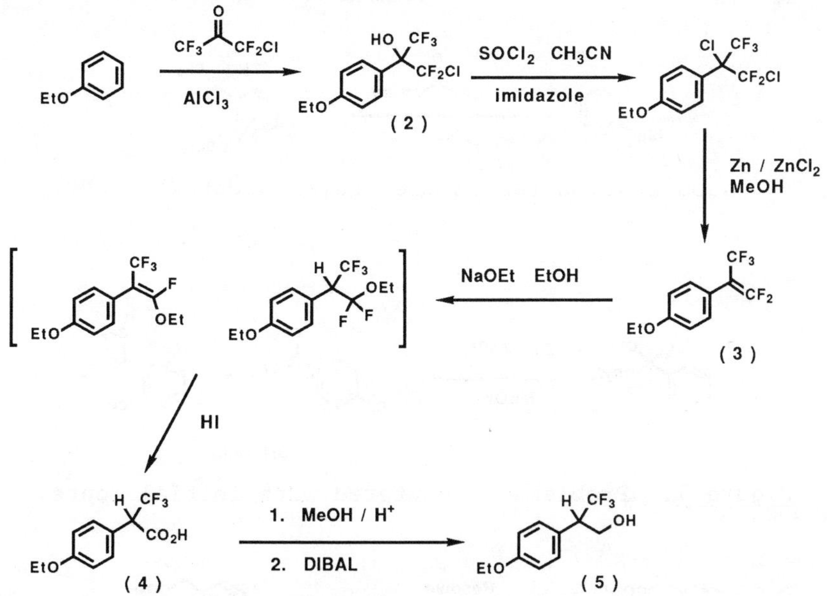

<u>Figure 6</u>. Synthesis of intermediate for preparation of targets (1a, 1b).

The initial synthetic route is shown in Figure 6, which is based on a literature synthesis of some anti-inflammatory α-trifluoromethyl phenylacetates[18].

Friedel Crafts reaction of chloropentafluoro-acetone with phenetole produced a tertiary alcohol (2), which was converted to the pentafluoropropene (3) by chlorination and dechlorination. Addition of ethoxide gave a mixture that was hydrolysed to the carboxylic acid (4), esterified and reduced to the alcohol (5). A phase-transfer catalysed coupling produced the target ethers (1).

This route works well with 4-ethoxy as a ring substituent and has been run through on scales >100g, but is, of course, useless for non-<u>para</u> directing substituents, and problems were encountered with t-butyl or chloro as shown in Figure 7. The starting fluorinated material is also expensive.

One attraction of this route was to use the pentafluoropropene (3) as a source of more highly fluorinated analogues, another was the ability to resolve a key intermediate (4).

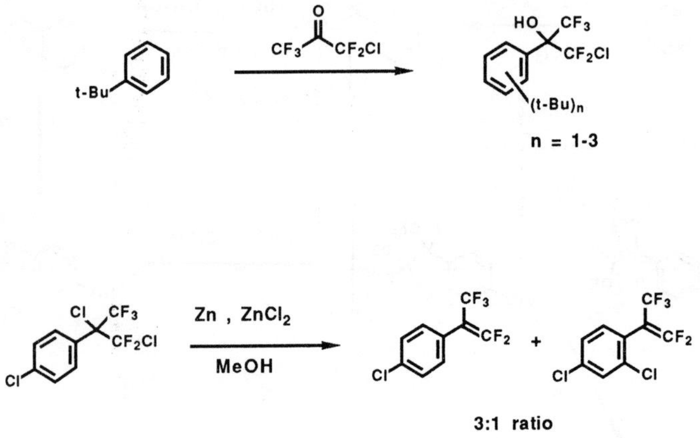

3:1 ratio

Figure 7. Problems encountered with initial route.

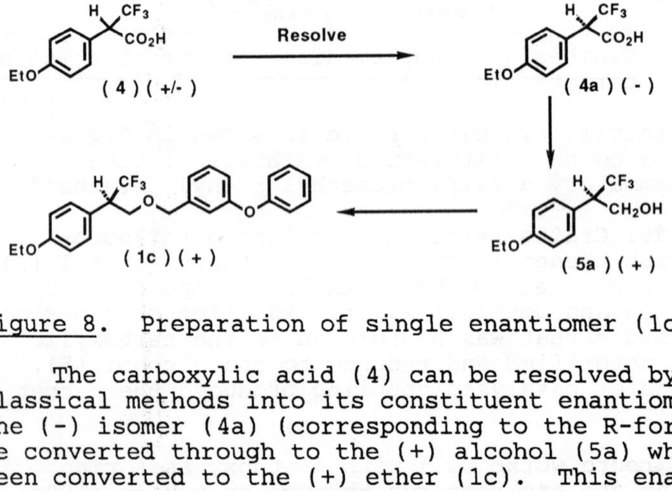

Figure 8. Preparation of single enantiomer (1c).

The carboxylic acid (4) can be resolved by classical methods into its constituent enantiomers[19]. The (-) isomer (4a) (corresponding to the R-form) can be converted through to the (+) alcohol (5a) which has been converted to the (+) ether (1c). This enantiomer has been shown by bioassays to contain practically all the insecticidal activity, while the corresponding (-) isomer is of very low activity. Thus fully resolved compounds will be about twice as active as the racemates.

This result apparently agreed with our view that only 1 methyl group (or equivalent) is important for activity. However, the absolute configuration (R) (assigned by Sumitomo[19]) is opposite to that predicted by our model. This is an intriguing result!

With such active lead compounds in hand a robust route was required that would be flexible enough to allow variation in all parts of the molecule, preferably incorporating the fluoroalkyl group without resort to drastic fluorination conditions or requiring expensive intermediates.

Improved Route to CF_3/H ethers e.g. (1)

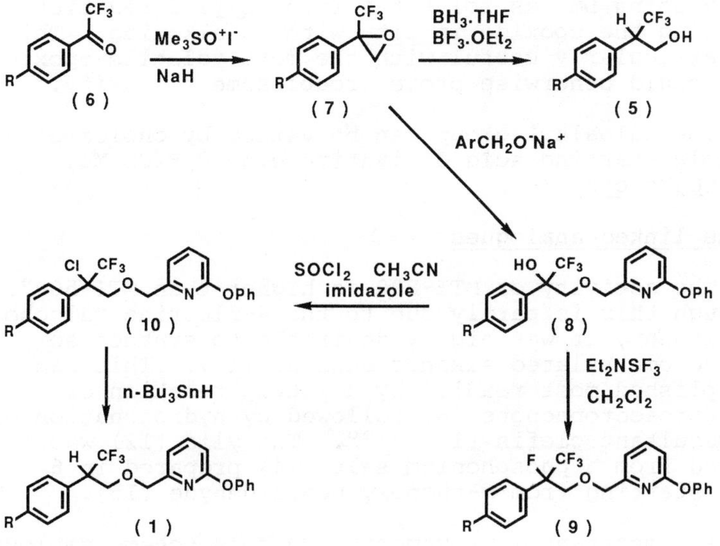

Figure 9.   Synthesis of ethers via trifluoroacetophenone intermediates.

Trifluoroacetophenones (6) were identified as important intermediates. They can be prepared[15] by reaction of a Grignard or aryl lithium species with trifluoroacetic acid (or esters) or trifluoroacetic anhydride. This strategy allows easy variation of the "acid end" ring substituents. Reaction with the ylid derived from trimethylsulphoxonium iodide gives the epoxide (7) which can be reduced selectively to the alcohol intermediate (5) for synthesis of targets by phase-transfer catalysed ether coupling.

Alternatively the epoxide (7) can be ring opened
with an alkoxide, derived from a pyrethroid alcohol, to
produce a tertiary alcohol (8) which allows some
variations α to the CF_3 to be explored (for example
F (9), Cl (10)). Figure 9 shows the phenoxypyridyl
analogues as an example of the easy variation of the
alcohol for SAR purposes. Tributyl tin hydride
reduction of the chloro compounds (10) provides an
additional route to the original lead compounds (1).
Direct conversion of (6) to (8) can be achieved in one
pot by using DMF as solvent for the S-ylid reaction and
quenching the epoxide in situ with an alkoxide. This
was particularly useful with the more volatile epoxides
which could otherwise prove troublesome to purify.

The haloalkyl group can be varied by choice of a
suitable starting acid derivative e.g. $C_2F_5CO_2Me$,
CF_2ClCO_2Me etc.

Alkane linked analogues

The activity of MTI-800 is higher than MTI-500[9].
Although this is partly due to the 4-fluorine "alcohol"
substituent, it was highly desirable to synthesise
trifluoromethylated alkanes such as (11). This was
accomplished most readily by a Wittig reaction of
trifluoroacetophenone (6) followed by hydrogenation of
the resultant olefin (13)[16,20]. The ylid (12) was
derived from a phosphonium salt (14) prepared in 6
steps starting from m-phenoxy benzaldehyde (15).

An important improvement[21] on this scheme employs
allylic phosphonium salts, (e.g. (18)), prepared in
only 3 steps from (15) by reaction with vinyl magnesium
bromide, treatment with HCl or HBr to produce the
rearranged allylic halide, and finally reaction with
Ph_3P. (Phosphonate derivatives were also made by an
Arbusov reaction). These allylic ylids (16) produce
diene intermediates (17) which are then hydrogenated to
the target alkanes (11).

Other routes to alkanes are shown in Figure 12. The
addition of an alkyl lithium (19) to (6) produces a
tertiary alcohol (20) - again allowing exploration of
substitution α to the CF_3, as well as dehydration to
the alkene. The other strategies remove the need to
synthesise a 3-carbon phosphonium salt, such as (14) or
(18), by building up the chain from the "acid" end,
followed by a subsequent Wittig using a reagent (21)
derived from e.g. 4-fluoro 3-phenoxybenzyl bromide.

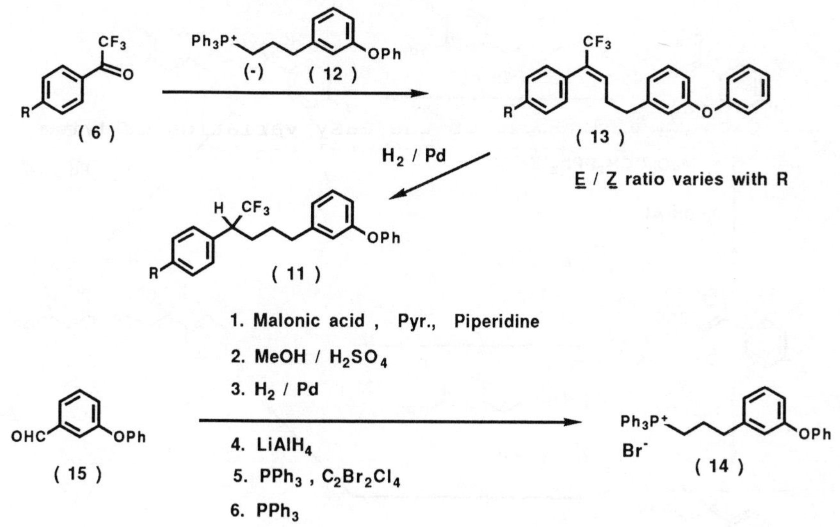

Figure 10. Synthesis of alkane analogues.

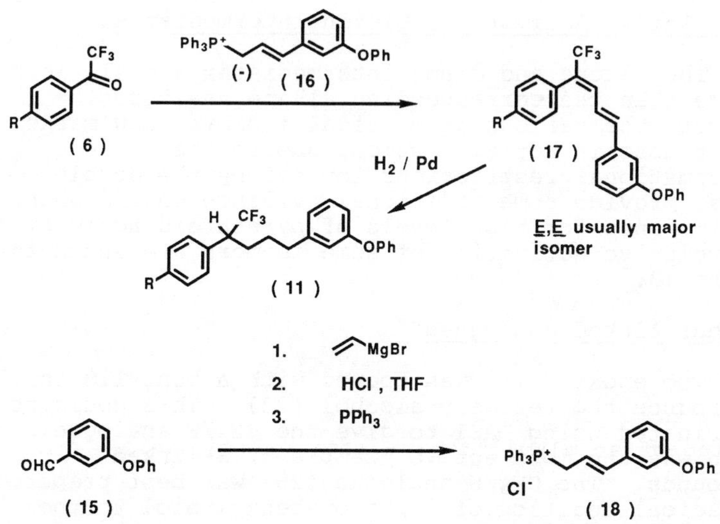

Figure 11. Improved route via allylic phosphonium salt

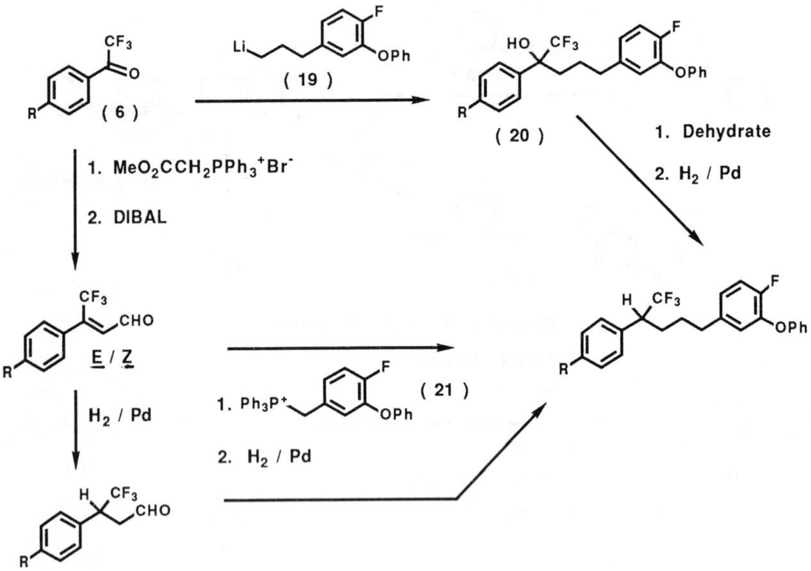

Figure 12.   Alternative routes to Alkanes.

Insecticidal Activity of Olefin Intermediates

The alkene and diene intermediates are all less
active than the corresponding alkane analogues.
However, the various geometrical isomers show marked
differences in activity which, due to the
conformational restrictions imposed by the double
bonds, provide some additional evidence on the shape
requirements for high levels of pyrethroid activity[12].
The relative activities of some isomers are shown in
Figure 13.

Sulphur linked analogues[22]

The epoxide (7) was opened with a benzylic thiol
to produce the tertiary alcohol (22). This could be
fluorinated using DAST to give the CF$_3$/F analogue (23)
which was the most active example of S-linked
compounds. The CF$_3$/H analogue (25) was best prepared
by radical addition of 3-phenoxybenzylthiol to the
styrene (24) although the yield was poor. This
compound proved considerably less potent than the ether
or alkane linked types.

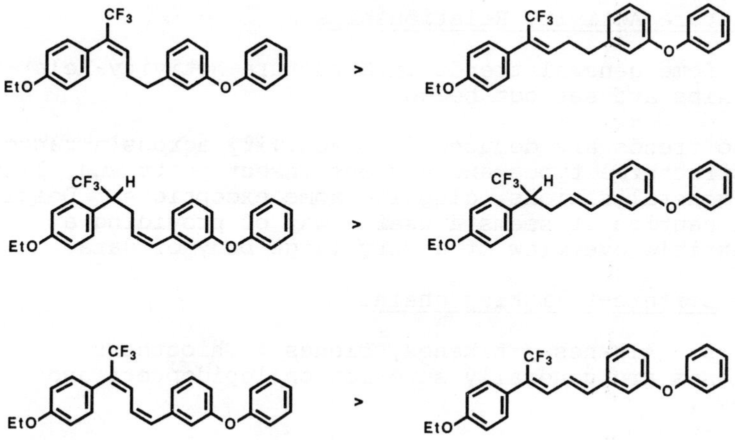

Figure 13. Relative activities of olefin and diene intermediates.

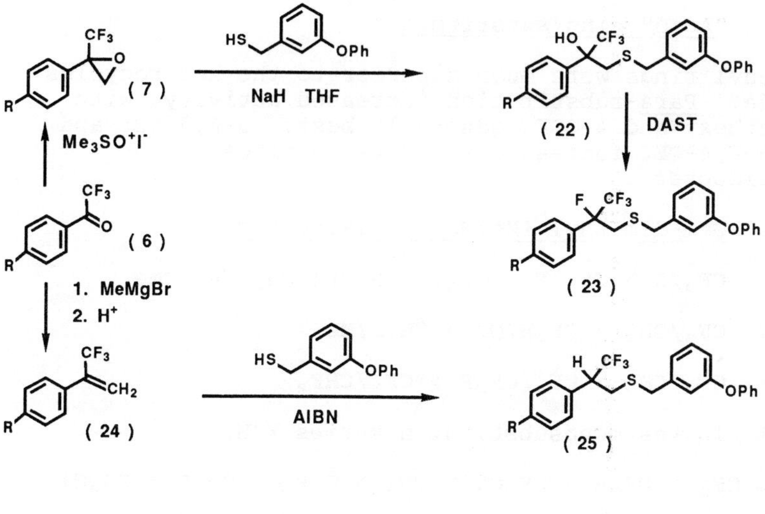

Figure 14. Synthesis of S-linked analogues.

Structure Activity Relationships

    Some general trends in Structure-Activity-Relat-
ionships are set out below.

These trends are deduced from activity across a range
of structural types and various insect pests and, like
all generalisations, disguise some exceptions. Despite
this caution it seems a useful way of providing a
digestible overview of a very large body of data.

a)   Different linking chains

Ethers > Alkanes > Alkenes, Dienes > Thioethers
(Alkanes are generally superior on lepidoptera and
mites).

b)   Alcohols

m-Phenoxybenzyl proved superior to all other alcohols
made (e.g. phenoxypyridyl; 2 methyl 3-phenylbenzyl;
fluorinated benzyls etc). The "Bayer alcohol"
(4-fluoro 3-phenoxybenzyl) generally provided the most
active analogues.

c)   "Acid" ring substituents

Phenyl rings were much superior to the heterocycles
made. Para-substitution increased activity, with
4-ethoxy and 4-OCF$_3$ generally best. 3-F,4-OEt and
3,5-F$_2$4-OEt increased activity on mites and
lepidoptera.

d)   Gem-dimethyl (Me/Me) replacements

i)   CF$_3$/H > CF$_3$/F > CF$_3$/Cl > CF$_3$/OH, CF$_3$/OMe

ii)   CF$_3$/CH$_3$ > CF$_2$H/Me > CH$_2$F/Me

iii) CF$_3$/CH$_3$ > CF$_3$/CH$_2$F > CF$_3$/CHF$_2$

iv)   In the monosubstituted series X/H,

X = CF$_3$ > CF$_2$H > CF$_2$Cl > CH$_3$ > C$_2$F$_5$ > CH$_2$F > CH$_2$Cl

The most striking of these trends is the effect of
increasing fluorine content of the gem-dimethyl group
(d ii) and iii)). Substitution of F for H in one

methyl group increases activity (CF_3/CH_3 > CH_3/CH_3) yet the more heavily fluorinated compounds (e.g. CF_3/CHF_2) are very weakly active. This suggests a beneficial receptor binding interaction for one fluorinated methyl group, but a detrimental interaction for a second fluorinated methyl group.

This probably correlates with the observation that in CF_3/H analogues, essentially all the activity resides in a single isomer. Although we have not resolved the CF_3/CH_3 types, we would predict a similar result. This dramatic difference is a good example of the difficulty of predicting sites where F for H substitution will increase activity. Given the current absence of any detailed knowledge at the molecular level of the pyrethroid receptor, the truly rational design of more active molecules remains a distant possibility.

Acknowledgements

The work described briefly in this presentation has involved a large number of people. The most important contributions to the chemistry have been made by A. J. Whittle, R. A. E. Carr, P. D. Bentley, L. Powell, F. J. Tierney, N. C. Sillars, P. Schofield, B. P. Shah, N. J. Bettesworth and M. D. Pearson. Prof. R. A. Raphael provided an excellent idea to facilitate synthesis of alkane linked analogues, and D. J. Tapolczay has coordinated large-scale synthesis. Biological testing has been supervised by J. C. Entwistle, R. F. S. Gordon and M. D. Collins.

I would like to thank everybody involved in this project for all their hard work.

Summary

A series of fluorinated analogues of the non-ester pyrethroids MTI-500 and MTI-800 have been described. Particularly high activity has been found with CF_3/H replacements for Me/Me. These can be prepared by short, efficient routes - incorporating the fluorine easily from trifluoroacetophenone intermediates. The compounds possess a single asymmetric centre; essentially all the insecticidal activity resides in one enantiomer.

Toxicity to mammals is, in general, exceptionally low with compounds of this type. The high level of activity demonstrated on the major insect pests of rice coupled with favourable toxicity to fish demonstrates the potential for this series to produce a commercially attractive rice insecticide.

REFERENCES

1.   M. Elliott, N. F. Janes, Chem.Soc.Rev., 1978, 7, 473.
2.   P. D. Bentley et al., Pestic.Sci, 1980, 11, 156.
3.   A. R. Jutsum et al."Proceedings of the British Crop Protection Conference", 1984 Vol.2 p421. M. J. Robson et al. ibid Vol. 3 p853.
4.   E. McDonald et al. "Proceedings of the British Crop Protection Conference" 1986, Vol. 1, p199. A. R. Jutsum et al. ibid, p97.
5.   J. F. Engel et al., "Proceedings of the Fifth International Congress of Pesticide Chemistry", ed. J. Miyamoto, P. C. Kearney, Kyoto, Japan 1982, Pergammon Press NY, 1983, Vol.1, p101.
6.   4'-Halogen e.g. Sumitomo, Patent Application J55073649, Dutch Patent NL 7908503.
7.   Alpha-ethynyl e.g. Sumitomo, Patent Application EP61713.
8.   "HX-addition" e.g. Sumitomo, Patent Application, EP44139.
9.   T. Udagawa et al. in "Recent Advances in the Chemistry of Insect Control" ed. N. F. Janes, 1985 (RSC Special Publication No. 53), P192.
10. M. Elliott et al. (NRDC), Patent Application BE902147.
11. G. Holan et al. (CSIRO), Patent Application WO8401-147.
12. Unpublished work by R. Taylor, ICI Agrochemicals (in part S. A. Lambros).
13. See for example A. J. Whittle et al. Patent Applications GB2184439, EP266891, GB2205096.
14. Lectures by C. Swithenbank given at Neurotox '85, Bath, England, 1985 and the SCI Meeting "Fluorine Containing Pesticides", London, England 21/10/85.
15. M. J. Bushell, A. J. Whittle, R. A. E. Carr, Patent Application GB2178739 (ICI).
16. H. Francke et al., Patent Application EP233834 (Schering).
17   K. Tsushima et al. Patent Application EP240978 (Sumitomo).

18. W. J. Middleton, E. M. Bingham, J.Fluorine Chem. 1983, 22, 561.
19. K. Tsushima et al., Agric.Biol.Chem., 1988, 52, 1323.
20. M. J. Bushell, R. A. E. Carr, Patent Application GB2187452 (ICI).
21. M. J. Bushell, R. A. Raphael, Patent Application EP279531 (ICI).
22. R. A. E. Carr, M. J. Bushell, Patent Application, EP273549 (ICI).

Silane Analogs of MTI-800: Biology and Chemistry

Scott McN. Sieburth, Sarah Y. Lin, John F. Engel, Jane A. Greenblatt, Susan E. Burkart, and Derek W. Gammon

AGRICULTURAL CHEMICAL GROUP, FMC CORPORATION, BOX 8, PRINCETON, NEW JERSEY 08543, USA

Silicon, the second most abundant element, plays an important role in both the polymer and electronic industries because of the special properties of siloxanes and silicon crystals. A significant role for silicon in agro-chemicals, however, has only recently appeared with the discovery of the fungicide DPX 6573,[1] the first silane to demonstrate a commercially viable level of pesticidal activity. Information concerning the metabolic and environmental stability of organosilanes is therefore very limited.[2]

Examples of silicon in insecticides are sparse. The earliest entry, a carbamate,[3] was inactive without pretreatment of the insect with piperonyl butoxide (PBO, an inhibitor of mixed function oxidase enzymes). A silane analog of DDT[4] proved to be inactive even with PBO pretreatment. The first organosilanes which were insecticidal without PBO, analogs of the juvenile hormones,[5] were reported in 1983.

We,[6] and others,[7] recognized the opportunity to test organosilane analogs of the pyrethroids which was presented by the development of ethofenprox (1) and MTI-800 (2).[8] The quarternary carbon, at the center of the (presumably) critical gem-dimethyl group,[9] seemed a particularly intriguing position to introduce silicon and to observe the biological effect resulting from the change in size that such a substitution would introduce. Compounds (1) and (2) were the first pyrethroid structures which allowed incorporation of silicon into this gem-dimethyl group: previous pyrethroids contained this unit as part of a three-membered ring and/or adjacent to a carbonyl group, features incompatible with organosilane stability.[10]

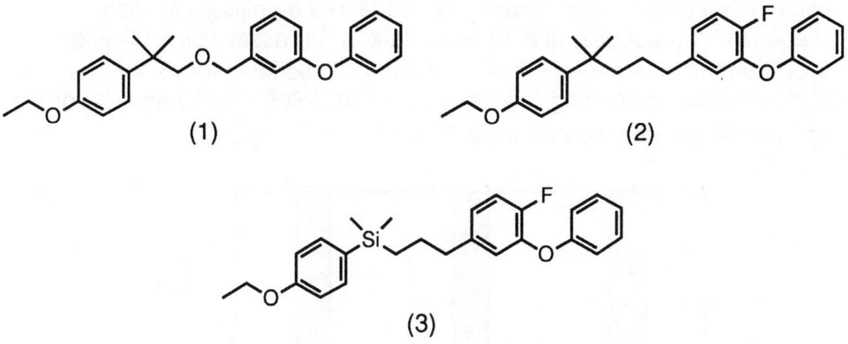

Figure 1. Ethofenprox (1), MTI-800 (2) and silane (3).

Insecticidal Activity

Topical testing of (2) and (3) on cabbage looper (*Trichoplusia ni*) and American cockroach (*Periplaneta americana*) demonstrated a similar magnitude of insecticidal activity for the two compounds (Table 1). Pretreatment of the insects with PBO resulted in an equal degree of synergism for the two compounds on cabbage looper, and no change in LD_{50} on the cockroach.

A comparison of silane (3), ethofenprox (1) and MTI-800 (2) in a foliar assay against representative Coleoptera, Lepidoptera and Homoptera is shown in Figure 2. The silane has a potency of 0.2-0.4 relative to the carbon analog (2), and is more active than (1) by a factor of 1.3-2.6. This is, therefore, the first silane to demonstrate a commercial level of insecticidal activity. The silane analogous to (1) had a relative potency of 0.2-0.5.

Fish Toxicity

The low fish toxicity of the lipophilic, non-ester pyrethroids such as (1) and (2) is a significant advance over the natural pyrethrins and other

Table 1

Topical LD_{50} values, in µg g^{-1} at 48 h

	Cabbage Looper		American Cockroach	
Compound	alone	PBO	alone	PBO
(2)	0.76	0.52	0.3	0.3
(3)	2.9	1.8	0.5	0.5

commercial pyrethroids. Silane (3) was tested on bluegill sunfish (*Lepomis macrochirus*) in a 48 hour static water assay using (2) and cypermethrin as standards.[9] No toxicity could be seen for (3) even at the highest aquarium loading rate of 50 ppm! MTI-800 (2) had an LC_{50} of 3 ppm, whereas the value for cypermethrin was 0.1 ppb.

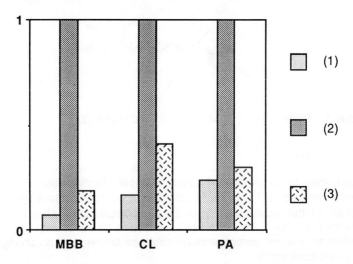

MBB = Mexican bean beetle   (*Epilachna varivestis*)
CL = cabbage looper   (*Trichoplusia ni*)
PA = pea aphid   (*Acyrthosiphon pisum*)

Figure 2. Relative potencies of (1), (2) and (3) in a foliar assay against three insect species.

The relative fish toxicity of insecticides has been rationalized as being due to the efficiency of gill uptake relative to compound lipophilicity.[9,11] Gill uptake efficiency is optimal for compounds with log P values in the range of 3 to 6 (Figure 3). Clearly, (2) and (3) are well outside this range.

Mode of Action

The mode of action of silane (3) was studied using electrophysiological techniques *in vitro* and *in vivo*.[9] Recording from the cercal nerves of the American cockroach 24 hours after topical application demonstrated only Type I-like effects, i.e. repetitive firing of the the cercal sensory nerve following stimulation. Such discharges were observed at doses up to 40 times the LD_{50}, a response which indicates a Type I pyrethroid-like action.

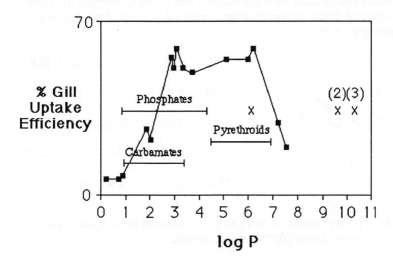

Figure 3. The relationship between log P and gill uptake efficiency (data from reference 11). Ranges of typical log P values for the commercial members of the three major insecticide classes are also shown. The log P values for (2), (3) and cypermethrin (unmarked X) were calculated using the program CLOGP (reference 9, 12).

Structure-Activity Relationships

Three regions of the structure of (3) were studied to determine their effect on insecticidal activity:[13] the substituent *para* to the silicon, the phenoxyphenyl group and one of the methyl groups on silicon.

Para Substituent. Figure 4 lists the *para* substituents which were prepared, in addition to ethoxy, in (3). QSAR analysis of the insecticidal activity of the monosubstituted set indicated that molar refractivity (MR) of the substituent was the primary contributing factor.[9]

Phenoxyphenyl Group. Replacement of the phenoxyphenyl with drastically different groups which are associated with pyrethroid alcohols such as the 2-methyl-3-biphenyl[14] or 4-methyl-2,3,5,6-tetrafluorophenyl[15] resulted in inactive structures. Within the phenoxyphenyl motif (Figure 5), bridging the two phenyl rings with carbon, nitrogen or oxygen yielded active compounds in each case, with the oxygen and carbonyl bridges the most active. None of these, however, approached the activity of the 4-fluoro-3-phenoxyphenyl in (3).

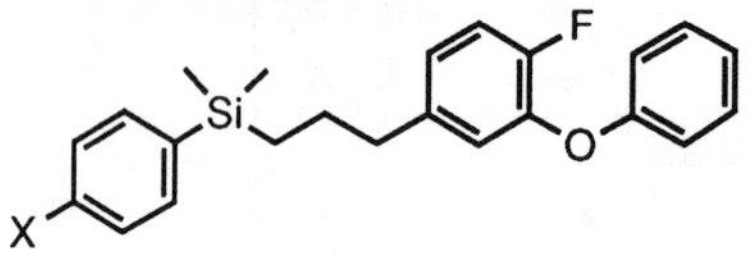

X	R.P.	X	R.P.
H	0.01	CF$_3$	0.03
Cl	0.05	(CH$_3$)$_2$CHO	0.006
CH$_3$	0.10	3,4-OCH$_2$O	0.31
CH$_3$O	0.15	3,4-(CH)$_4$	0.01

(4)

Figure 4. Potency values, relative to (3), in foliar assays against Mexican bean beetle (*Epilachna varivestis*).

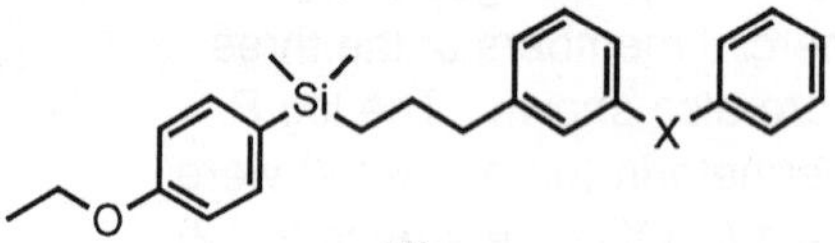

X	MBB	SAW
O	0.08	0.09
CO	0.22	0.05
CH$_2$	0.02	0.02
NH	0.01	0.02

(5)

Figure 5. Potency values, relative to (3), in foliar assays against Mexican bean beetle (*Epilachna varivestis*) and southern armyworm (*Spodoptera eridania*).

Replacement of a Methyl Group. One of the methyl groups of (3) was changed to vinyl (6a), ethyl (6b) and cyclopropyl (6c), and the biological activity for these is shown in Figure 6. This set of increasingly larger groups resulted in a rapidly decreasing level of insecticidal activity. This apparent relationship between size and activity would be anticipated based on the assumed correspondence of these methyl groups with the gem-dimethyl group of the cyclopropyl and isopropyl ester pyrethroids which are very sensitive to change.

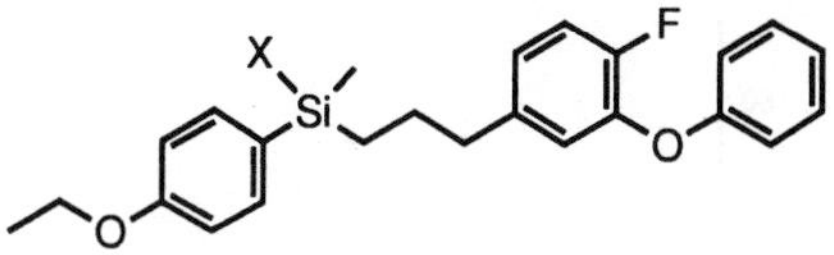

	X	R.P.
(6a)	$CH=CH_2$	0.012
(6b)	C_2H_5	0.009
(6c)	$\underline{c}$-C_3H_5	inactive
(6d)	H	0.12
(6e)	OH	inactive

Figure 6. Potency values, relative to (3), in foliar assays against Mexican bean beetle (*Epilachna varivestis*).

Replacement of the methyl group with a hydrogen would be anticipated to result in little or no activity due to the removal of a critical group and the reported *in vivo* lability of the silicon-hydrogen bond.[16] However, silane (6d) proved to be significantly active both in topical and foliar tests against cabbage looper, with an LD_{50} value of 10 µg g^{-1} topically. Pretreatment of the cabbage looper with a five fold dose of PBO before topical application of (6d) resulted in an LC_{50} value of 6.8 µg g^{-1}, a level of synergism (1.5x) equivalent to that found for silane (3) and MTI-800 (2). It is surprising that this allegedly labile functionality demonstrates so little synergism. The anticipated product of oxidative metabolism, silanol (6e), was inactive both in foliar tests and topically up to 10,000 µg g^{-1}. Apparently, *in vivo* oxidation of (6d) is much slower than the manifestations of its toxic effects.

<u>Field Activity</u>

Field trials results for (3) and relevant standards on four insect pests are shown in Figure 7.

The insecticidal activity of (3) was measured against two leafhopper species using 110 g ha^{-1}. Treatment of rice (Graph A) gave control of green leafhopper (*Nephotettix virescens*) equivalent to, or better than, an equal treatment with ethofenprox (1). Application of (3) on grapes gave excellent control of variegated leafhopper (*Erythoneura variabilis*) for at least 28 days, similar to the control shown by dimethoate at 2200 g ha^{-1} (Graph B).

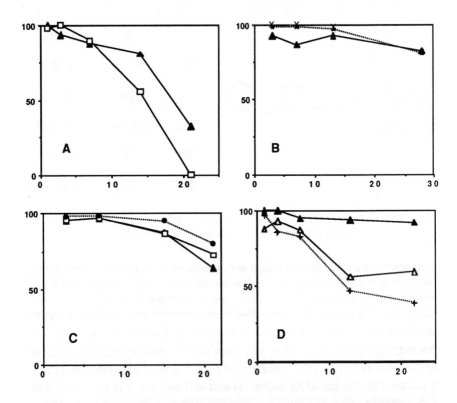

Figure 7. Field test results, in Percent Control (ordinate) <u>versus</u> Days
After Treatment (abscissa). Test compounds: ▲ (3) 110 g ha^{-1};
Δ (3) 55 g ha^{-1}; □ (1) 110 g ha^{-1}; • cypermethrin 44 g ha^{-1};
x dimethoate 2200 g ha^{-1}; + carbaryl 1100 g ha^{-1}.
A. Green leafhopper (*Nephotettix virescens*) on rice;
B. Varigated leafhopper (*Erythoneura variabilis*) on grapes;
C. Alfalfa weevil (*Hypera postica*) on alfalfa;
D. Gypsy moth (*Lymantria dispar*) on oak.

Both ethofenprox (1), at 110 g ha^{-1}, and cypermethrin, at 44 g ha^{-1}, were
controls in a test against alfalfa weevil (*H. postica*). Silane (3) at 110 g ha^{-1}
was as almost effective as (1) and Cypermethrin at 21 days (Graph C).

Gypsy moth (*L. dispar*) defoliates large areas of the northeastern United

States annually, with oak as its preferred host. Silane (3) at 110 g ha⁻¹ demonstrated nearly complete control to 22 days. At one half of this rate, (3) was superior for the control of gypsy moth to carbaryl at 1100 g ha⁻¹, which is a typical treatment for this pest (Graph D).

Conclusion

Silane (3) is an effective, broad spectrum insecticide with a very low toxicity to fish. Studies of the synergistic effect on toxicity of the oxidase inhibitor piperonyl butoxide for both cabbage looper and the American cockroach demonstrated that this metabolic detoxification pathway is no more important for (3) than for the carbon analog (2). Furthermore, the analog of (3) containing a silicon-hydrogen bond was only slightly affected by the synergistic effects of piperonyl butoxide, suggesting a broader role for silicon in insecticides than previously thought possible.

Results of field tests with (3) demonstrated good-to-excellent control of insects at low application rates, making it the first organosilane to possess a commercial level of insecticidal activity.

References

1. W.K. Moberg, G.S. Basarab, J. Cuomo and P.H. Liang. 'Synthesis and Chemistry of Agrochemicals', American Chemical Society, Washington, D.C., 1987, p.288.

2. R. Tacke and H. Linoh. 'The Chemistry of Organic Silicon Compounds'. S. Patai and Z. Rappoport, Eds., John Wiley & Sons Ltd., 1989, Vol. 2, Chapter 18, p.1143.

3. R.L. Metcalf and T.R. Fukuto. *J. Econ. Entomol.,* **1965**, *58*, 1151.

4. M.A.H. Fahmy, T.R.Fukuto, R.L. Metcalf and R.L. Holmstead. *J. Agric. Food Chem.,* **1973**, *21*, 585.

5. I. Ujvary, A. Kis-Tamas, L. Varjas and L. Novak. *Acta Chim. Hungarica,* **1983**, *113*, 165.

6. US Patent 4 709 068 (1987).

7. Y. Yamada, T. Yano and N. Itaya. *J. Pestic. Sci.,* **1987**, *12*, 683; Sumitomo, JP 60-123491 (1985); Katsuda, JP 61 087 687 (1986);

Shionogi, USP 4 663 314 (1986); Hoechst, EP 224 024 (1987); Mitsui Toatsu, JP 62 108 885 (1987).

8. T. Udagawa, S. Numata, K. Oda, S. Shiraishi, K. Kodaka and K. Nakatani. 'Recent Advances in the Chemistry of Insect Control'. N.F. Janes, Ed., The Royal Society of Chemistry, London, 1985, p.192.

9. S.McN. Sieburth, C.J. Manly and D.W. Gammon. *Pestic. Sci.*, **1989**, *24*, 0000.

10. I. Fleming. 'Comprehensive Organic Chemistry'. D. Barton and W.D. Ollis, Eds., John Wiley & Sons Ltd., New York, 1987, Vol. 3, p.539.

11. J. McKim, P. Schmieder and G. Veith. *Toxicology and Applied Pharmacology*, **1985**, *77*, 1.

12. A.J. Leo. 'QSAR and Strategies in the Design of Bioactive Compounds.' J.K. Seydel, Ed., VCH Publishers, Deerfield Beach, Florida, 1985, p.294.

13. S.McN. Sieburth, C.J. Langevine and D.M. Dardaris. *Pestic. Sci.*, **1989**, *24*, 0000.

14. J.F. Engel, E.L. Plummer, R.R. Stewart, W.A. Van Saun, R.E. Montgomery, P.A. Cruickshank, W.N. Harnish, A.A. Nethery and G.A. Crosby. 'Proceedings of the Fifth International Congress of Pesticide Chemistry'. J. Miyamoto and P.C. Kearney, Eds., Pergammon Press, New York, 1983, p.101.

15. A.R. Jutsum, R.F.S. Gordon and C.N.E. Ruscoe. 'Proc. Brit. Crop Protect. Conf. -- Pests and Diseases,' 1986, Vol. 1, 97.

16. R.J. Fessenden and R.A.Hartman. *J.Med. Chem.*, **1970**, *13*, 52.

Some Aspects of Synthesis and Structure-Activity in Insecticidal Lipid Amides

Robert J. Blade

WELLCOME RESEARCH LABORATORIES, BERKHAMSTED, HERTS. HP4 2DY, UK

INTRODUCTION

A diverse range of lipophilic amides occurring in plants of the <u>Compositae</u> and <u>Rutaceae</u> families exhibit a variety of biological activities; of particular interest are the many examples with insecticidal activity.[1] One of the first structures to be elucidated was that of the simple decadienamide pellitorine (1)[2]. Subsequently a number of more complex amides have been isolated and identified of which neoherculin (2)[2] and anacyclin (3)[2] are examples. The <u>Piper nigrum</u> plant has been a fertile source from which Miyakado's group have isolated insecticidal aromatic lipid amides such as pipercide (4) and guineensine[1].

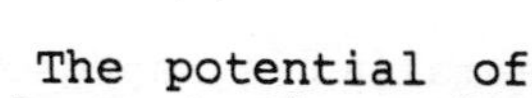

(1) 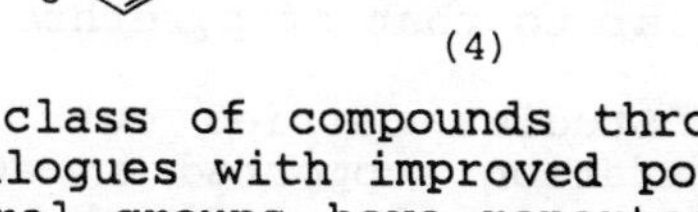   (3)

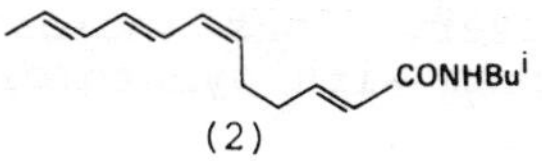

(2) 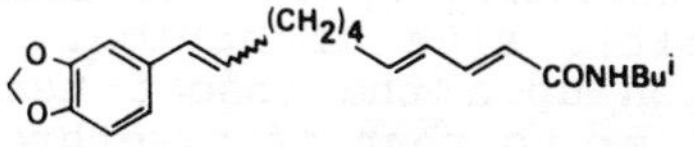   (4)

The potential of this class of compounds through development of synthetic analogues with improved potency is well recognised and several groups have reported on research into structural modifications. Our own studies started from the generalised active structure (5) which arose from a consideration of the insecticidal natural amides and we chose to investigate three classes of synthetic analogue as shown in Fig.1.

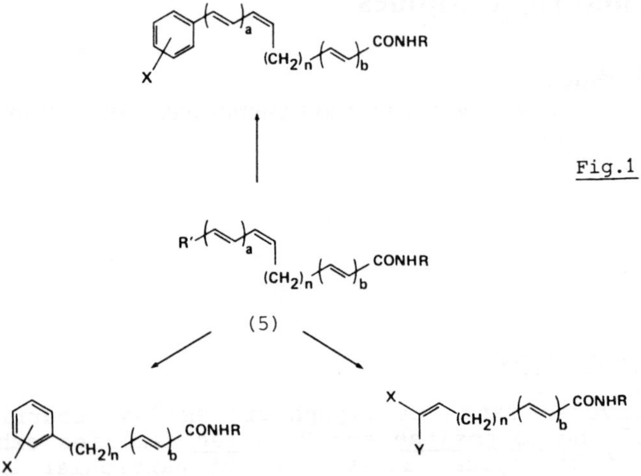

Fig.1

Arylpolyenamides and benzyloxyalkyldienamides

Initial studies led to the preparation of arylpolyenamides (6, 7 and 8) which showed some improvement in
potency against houseflies over the natural materials (M.
domestica) (Fig.2). The low levels of potency were off-
set however by the lack of cross-resistance to the kdr
resistant housefly strain (Fig.3). This lack of pyrethroid
cross resistance to lipid amides has been investigated in
detail by Elliott, Sawicki et al.[3]

The dependence of activity upon the number and
stereochemistry of the double bonds is suggestive of a
specific site of action. Early observations indicated
action upon the insect nervous system with symptomology
similar to that of pyrethroids.

Studies carried out in a mammalian synaptosomal
preparation supported an action upon voltage-dependent
sodium channels   (Fig 4a) in which release of a neurotransmitter precipitated by amides could be blocked by
tetrodotoxin. In an insect preparation (Fig 4b) the
amides did not produce direct depolarisation and the
effects upon veratridine induced depolarisation were
observed.[4] The amides inhibited the response in a manner
reflecting relative *in vivo* activities, whereas
deltamethrin enhanced the response.

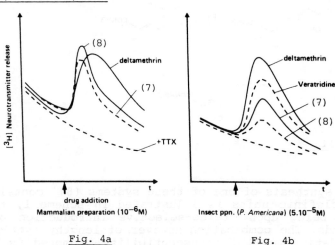

Topical activity vs housefly (*M. Domestica*)
(synergised; permethrin = 100)

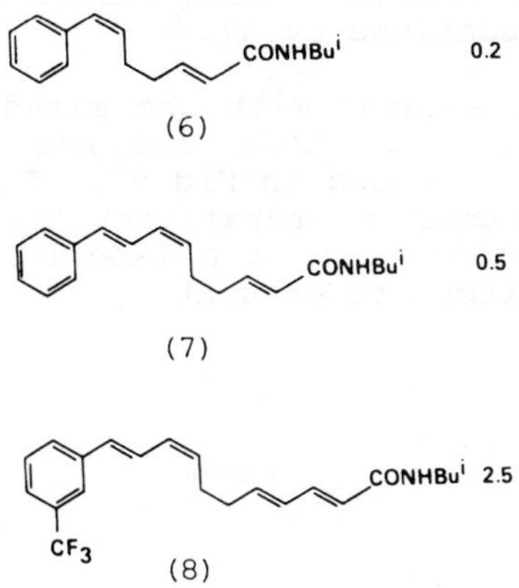

Activity vs kdr resistant houseflies

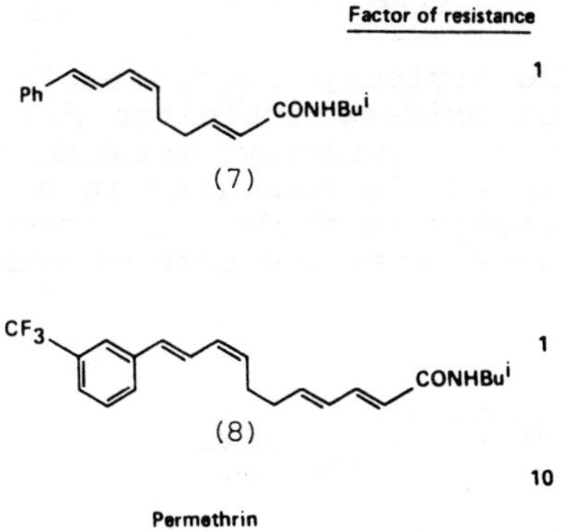

CONHBu^i 0.2

(6)

Factor of resistance

Ph   CONHBu^i 1

(7)

CONHBu^i 0.5

(7)

CF$_3$ CONHBu^i 1

(8)

CONHBu^i 2.5

CF$_3$ (8)

10

Permethrin

Fig. 2

Fig. 3

Synaptosomal Release experiments

[^{3}H] Neurotransmitter release

(8)

deltamethrin

(7)

+TTX

↑ drug addition
Mammalian preparation (10^{-6}M)

deltamethrin

Veratridine

(7)

(8)

Insect ppn. (*P. Americana*) (5.10^{-5}M)

Fig. 4a

Fig. 4b

These results would seem to reflect the behaviour against pyrethroid resistant houseflies and suggest a different site of action for amides in the insect sodium channel complex.   Some similar effects have been observed in binding studies conducted by Soderlund et al.[5]

The arylpolyenamides exhibit synergy with the mixed function oxidase inhibitor piperonyl butoxide and are sensitive to oxidative metabolism as shown in Fig 5[6].   The diene in (7) is converted in a mammalian preparation via the epoxides to diols.   In insects the use of a labelled probe shows that the same metabolites are present.

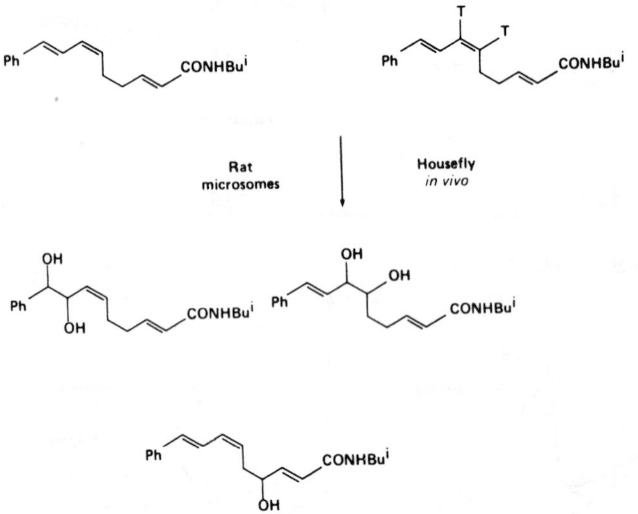

Fig. 5

The synthesis of one of these systems (8)[7] containing several olefinic units is illustrated in scheme 1, the emphasis being upon stereo-selective introduction of doublebonds. The combination however, of lengthy synthesis with oxidative metabolic susceptibility, prompted us to seek out a system in which the aryldiene fragment is replaced by a more accessible moiety.   The benzyl ethers shown in scheme 2 meet this requirement.

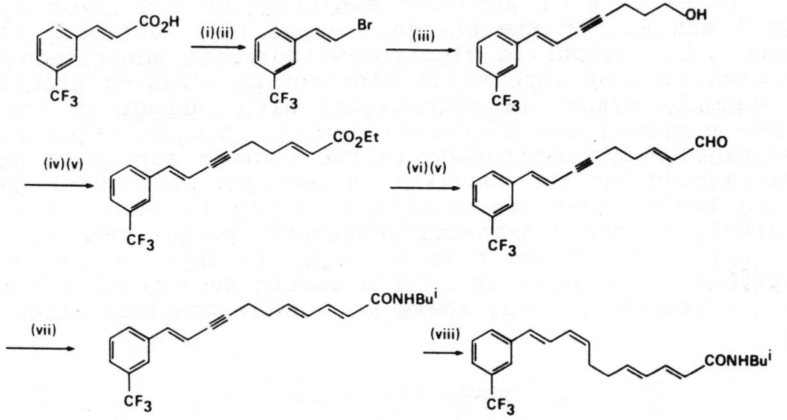

(i) $Br_2.CHCl_3$ (ii) Na_2CO_3 (iii) Pentyne—1—ol, $(Ph_3P)_2PdCl_2$—CuI,NHEt$_2$ (iv) PCC,CH_2Cl_2 (v) $Ph_3P:CH.CO_2Et$, CH_2Cl_2 (vi) $HAl(Bu^i)_2,CH_2Cl_2$ (vii) $Ph_3P:CH.CONBu^i$, MeOH (viii) Lindlar cat., EtOAc, H_2

Scheme 1

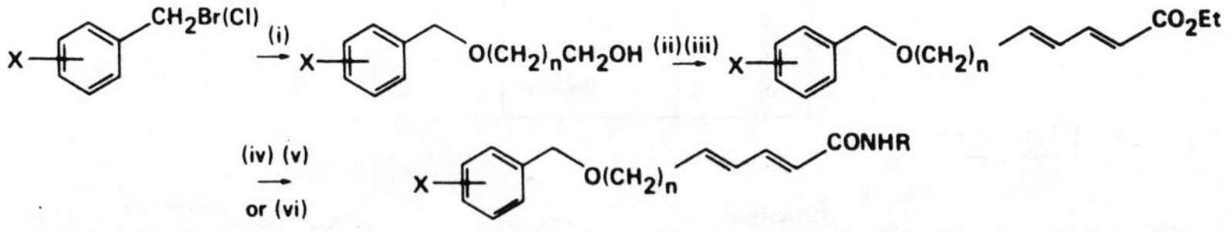

(i) Na $O(CH_2)_nCH_2OH$, PhMe (ii) DMSO—$(COCl)_2$, NEt$_3$, CH_2Cl_2 (iii) $(EtO)_2PO.CH_2CH:CH.CO_2Et$, LDA,THF
(iv) KOH,EtOH, H_2O (v) PhNH.P(O)Cl.OPh, NH$_2R$, NEt$_3$,CH_2Cl_2 (vi) Me_3Al, NH$_2R$, PhMe

Scheme 2

The relative ease of synthesis of these materials[8] rendered them an attractive series in which to evaluate the effects upon insecticidal activity of modification in the various regions of the molecule.

The overall length of the system is a key parameter and this is shown in Fig 6; the activity is assessed by comparison with the most active member of the group. In this case activity is optimised when the value of n is four or six, corresponding to an overall six or eight units between the aromatic nucleus and dienamide. These observations are in agreement with those made by other groups active in the area.[1,9]

The effects of aromatic substitution are shown in Figs 7 and 8. In the shorter series (n=4) activity is enhanced by lipophilic electron-withdrawing substituents. The inclusion of more polar electron-withdrawing groups, for example cyano, is detrimental with respect to the parent compound and electron-donating groups offer no advantage. Polysubstitution in the shorter series is not advantageous but the converse is the case with the longer series (n=6) where appreciable activity is found, particularly in the 3,5-bistrifluoromethylbenzyl ethers. It does appear therefore that activity in this region is sensitive to changes in both aromatic substitution and overall length and that there are subtle combined effects.

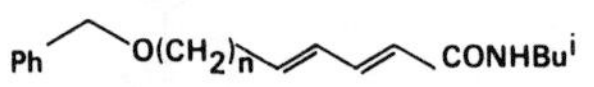

Ph—O(CH$_2$)$_n$—CONHBu^i

n	Housefly
2	0.1
3	0.1
4	0.4
5	0.2
6	1
7	0.4
8	0.25

Fig. 6

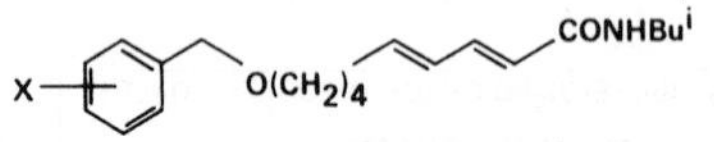

X—⬡—CH$_2$—O(CH$_2$)$_4$—CONHBu^i

X	Housefly (permethrin 100)
H	0.2
3–Cl	2
3–Br	2
3–F	1
3–CF$_3$	5
4–CF$_3$	1
3–CN	<0.1
4–OEt	0.2
3,4–methylenedioxy	0.6
3,5–(CF$_3$)$_2$	3
2,4–Cl$_2$	0.5

Fig. 7

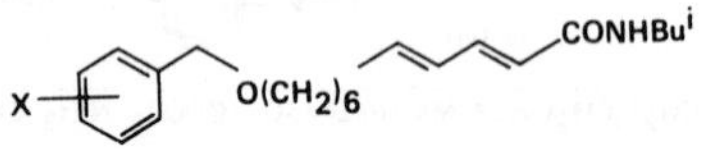

X—⬡—CH$_2$—O(CH$_2$)$_6$—CONHBu^i

X	Hf
H	0.6
3 – CF$_3$	2
4 – CF$_3$	1
3,5 – (CF$_3$)$_2$	28
2,4 – (CF$_3$)$_2$	9
3,5 – Cl$_2$	8
2,4 – Cl$_2$	5

Fig. 8

The key pharamacophore is the dienamide moiety and this is very sensitive to structural modification (Figs. 9 and 10). The stereochemistry is critical, the (2$\underline{E}$,4$\underline{E}$) dienamides being considerably more active than any of the other arrangements of double bonds. The geometry and degree of unsaturation afforded by acetylenic linkages are similarly disadvantageous although some activity is associated with 2-ene-4-ynamides.

Substitution of the diene brings about substantial changes in activity. In a series of methyl substituted compounds only the 3-methyldienamides retain appreciable activity. The latter position is itself sensitive to further modification which alters either the electronic or steric nature of the substituent.

The activities of these analogues are cited with respect to the most potent of the group (activity=1) and their syntheses are detailed in figs. 9 and 10.

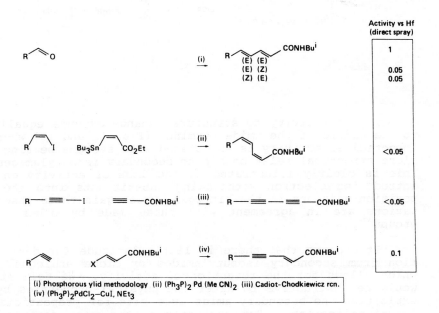

(i) Phosphorous ylid methodology (ii) (Ph$_3$P)$_2$ Pd (Me CN)$_2$ (iii) Cadiot-Chodkiewicz rcn.
(iv) (Ph$_3$P)$_2$PdCl$_2$–CuI, NEt$_3$

Fig. 9

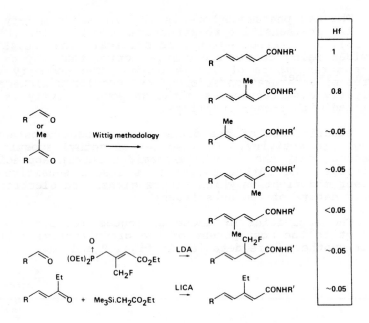

Fig. 10

The sensitivity to structural change extends equally to the nature of the amide terminus (Figs 11 and 12) where insecticidal activity is associated with systems bearing a close structural relationship to secondary isobutylamides. This is clearly illustrated by the loss of activity on introducing electron withdrawing substituents into the position adjacent to the nitrogen. Once again these observations[10] are in agreement with those made by other groups.

It appears that there is little latitude for deviation from secondary isobutylamides or close analogues which allows for the retention of activity. We felt it would be instructive to alter the physical-properties by exploiting the secondary amide as a site for propesticide functionalisation. Two synthetic routes were devised (Schemes 3 and 4), the first involving reaction of a dienoyl halide with a suitable lithiated intermediate and the second the silylated secondary amide.[11]

Using these approaches active compounds did emerge but they presented no advantages over the parent compounds. A particular group of propesticides are the unsaturated thioamides prepared by the route shown in scheme 5[12]; these showed somewhat reduced activity <u>in vivo</u> against houseflies when compared to the parent amides but much reduced <u>in vitro</u> potency.

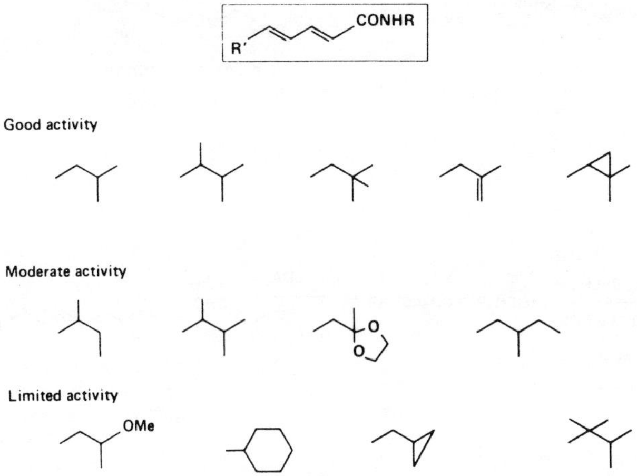

Good activity

Moderate activity

Limited activity

<u>Fig. 11</u>

Declining activity

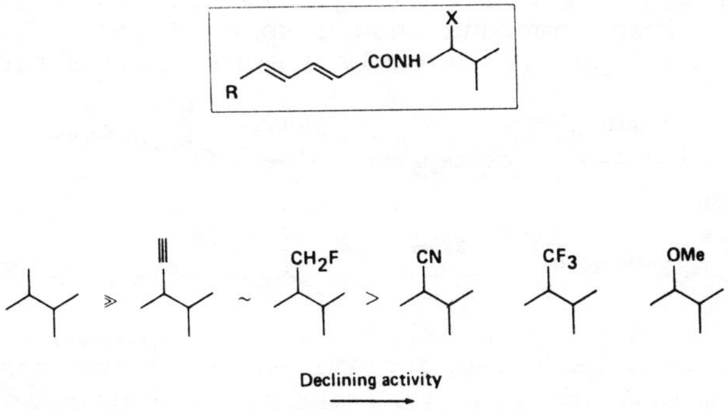

<u>Fig. 12</u>

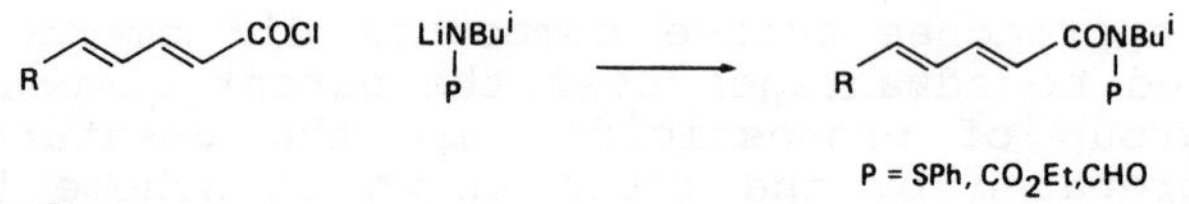

R⌒⌒COCl LiNBu^i ⟶ R⌒⌒CONBu^i
          | |
          P   P

P = SPh, CO$_2$Et, CHO

Scheme 3

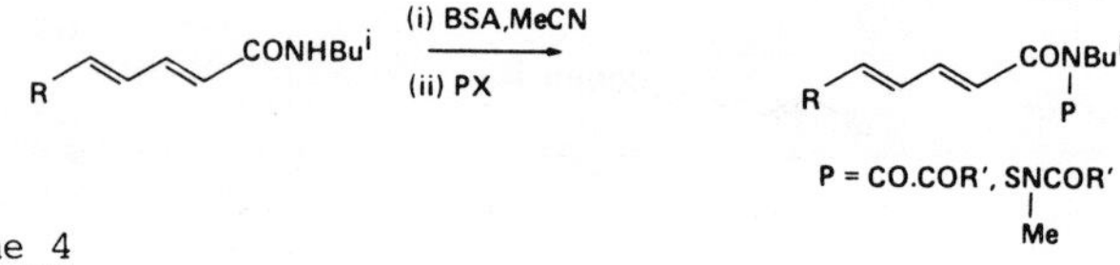

R⌒⌒CONHBu^i (i) BSA,MeCN ⟶ R⌒⌒CONBu^i
                (ii) PX |
                                             P

P = CO.COR', SNCOR'
                                             |
                                             Me
Scheme 4

(EtO)$_2$P(O)Me (i) BuLi ⟶ (OEt)$_2$P(O)CH$_2$CSNHBu^i (i) LDA ⟶ R⌒⌒CHO R⌒⌒C(=S)NHBu^i
                 (ii) Bu^{i}NCS   (ii) R

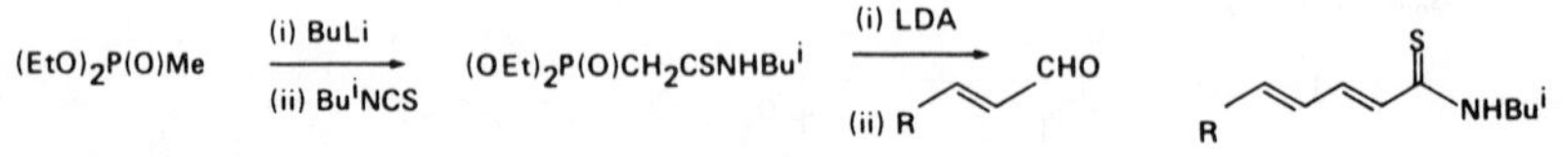

Scheme 5

Difluorovinylalkyldienamides

The second class of amide of interest are aliphatic systems bearing a halogenated olefinic terminus, as illustrated by compound (10) synthesised according to scheme 6[13]. This compound showed some marked differences in its behaviour <u>in vivo</u>, notably a more rapid action.

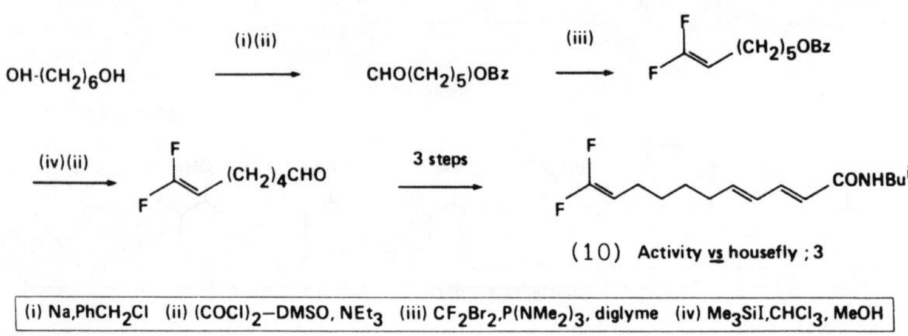

OH-(CH$_2$)$_6$OH (i)(ii) ⟶ CHO(CH$_2$)$_5$OBz (iii) ⟶ F$_2$C=CH(CH$_2$)$_5$OBz

(iv)(ii) ⟶ F$_2$C=CH(CH$_2$)$_4$CHO 3 steps ⟶ F$_2$C=CH(CH$_2$)$_3$CH=CH-CH=CH-CONHBu^i

(10) Activity <u>vs</u> housefly ; 3

(i) Na,PhCH$_2$Cl (ii) (COCl)$_2$–DMSO, NEt$_3$ (iii) CF$_2$Br$_2$,P(NMe$_2$)$_3$, diglyme (iv) Me$_3$SiI,CHCl$_3$, MeOH

Scheme 6

A neurophysiological evaluation [14,15] revealed some subtle differences in response on comparison with aromatic lipid amides. Changes in three parameters were monitored, the extent of depolarising after-potential, depolarisation of resting-potential and depression of action-potential (Fig 13a). The results are shown schematically in Fig 13b. A representative benzyl ether (9) (cf.Fig 7) affected only two of these parameters as did deltamethrin at a lower concentration. The aliphatic amide (10) showed a pronounced depolarising after-potential and little depression of action-potential, a profile more reminiscent of a type 1 pyrethroid such as S-bioallethrin. (Fig 13b)

Mannitol gap preparation in cockroaches (*P. Americana*)

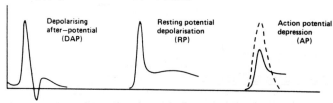

Fig. 13a

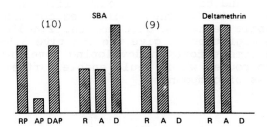

Fig. 13b

    The pharmacokinetics of compound (10) are different to those of the aromatic system (9) and this is reflected by the more rapid penetration of the former in houseflies. (Fig 14) [16]. These results indicate that enhancement of potency in vivo depends upon both pharmacodynamic and pharmacokinetic factors.

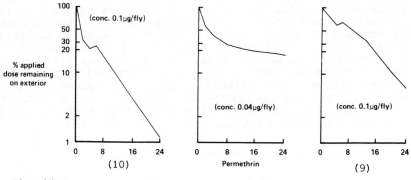

Fig. 14

Arylalkyldienamides

A major class of interest are the arylalkyldienamides (11). These have also been the subject of investigations by Elliott and his group as exemplified by the bromonaph-thyl system (12)[3]. These molecules contain a base sensi-tive, labile methylene unit and the 2,4-diene fragment is prone to migration to give the inactive 3,5-dienamides. We invested some effort in developing novel non-Wittig type synthetic routes which would enable further evaluation of this class of compounds.

Our first route[17] (Scheme 7) made use of the molyb-denum catalysed elimination of acetates, developed by Trost *et al.*[18], in a synthesis of piperovatine. Reaction of a terminal aldehyde with an arylsulphinyl acetate followed by 2,3-sigmatropic rearrangement afforded a pre-cursor acetate which upon elimination gave the dienoate without double-bond migration. This was converted, under non-basic conditions, to the final target.

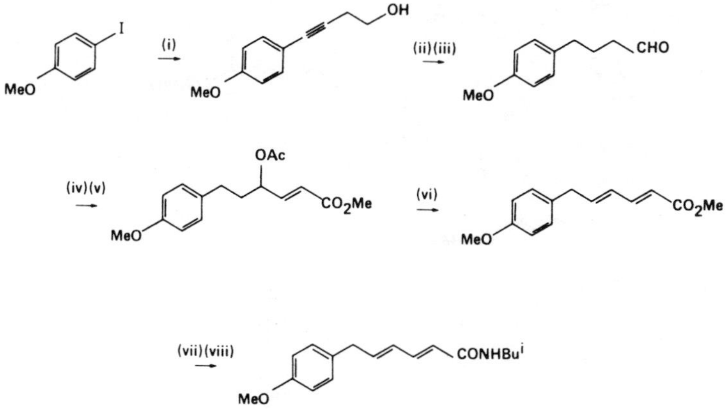

(i) (Ph$_3$P)$_2$ PdCl$_2$–CuI, Butynol (ii) H$_2$, Pd/C (iii)(COCl)$_2$–DMSO, NEt$_3$ (iv) Cl.C$_6$H$_4$S(O).CH$_2$CO$_2$Me, piperidine, MeCN (v) Ac$_2$O, DMAP, pyr. (vi) Mo (CO)$_6$, BSA, PhMe (vii) HCl, H$_2$O, dioxan (viii) PhNH.P(O).Cl.OPh,NH$_2$Bu^i,NEt$_3$.

Scheme 7

A convergent approach of particular use for systems with an electron-deficient aryl nucleus is shown in scheme 8 and is based upon methodology by Tamura *et al.*[19] The key step in this sequence is the Ene type reaction of a terminal olefin with an intermediate sulphonium salt derived from a sulphinyl acetamide. The oxidation and subsequent elimination of the adduct gives the 2,4-dienamides.

A third route of interest is a versatile convergent synthesis developed in collaboration with Prof. Crombie and his group[20]. This makes use of the palladium [0] catalysed coupling of an iodoacrylamide with a vinyl zirconocene (scheme 9) to give, in this case, piperovatine with good regio- and stereochemical control.

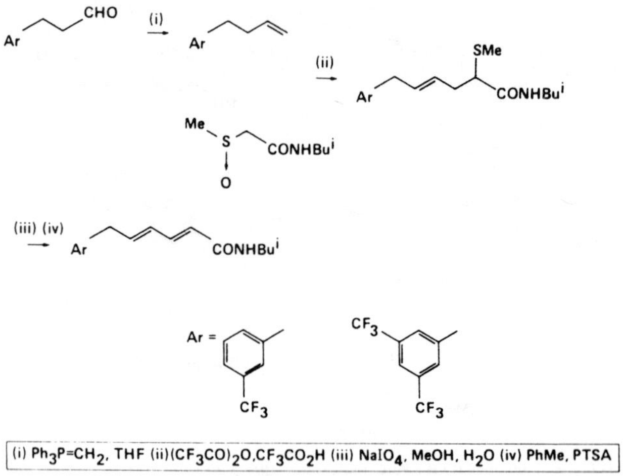
CHO
Ar
(i)
Ar
SMe
(ii)
Ar
CONHBu^i
Me
S
CONHBu^i
O
(iii) (iv)
Ar
CONHBu^i
Ar =
CF_3
CF_3
CF_3
(i) $Ph_3P=CH_2$, THF (ii) $(CF_3CO)_2O$, CF_3CO_2H (iii) $NaIO_4$, MeOH, H_2O (iv) PhMe, PTSA

Scheme 8

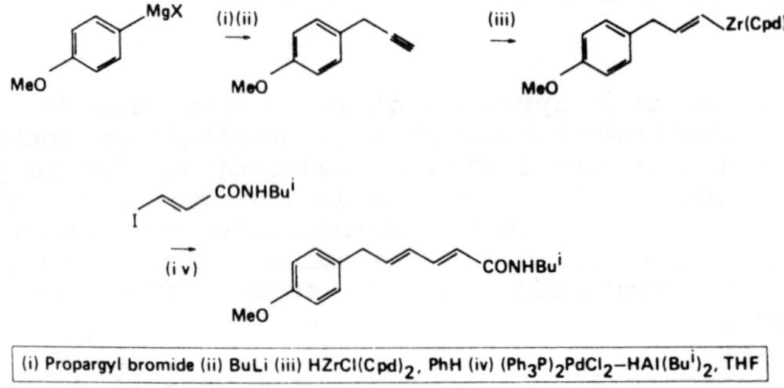
MgX
MeO
(i)(ii)
MeO
(iii)
Zr(Cpd)$_2$
MeO
CONHBu^i
I
(iv)
CONHBu^i
MeO
(i) Propargyl bromide (ii) BuLi (iii) $HZrCl(Cpd)_2$, PhH (iv) $(Ph_3P)_2PdCl_2$–$HAl(Bu^i)_2$, THF

Scheme 9

We felt however that a better strategy would be to identify active analogues from which the labile methylene unit were absent; such a group of compounds are the bicyclic systems (13).[21] We found that both tetrahydronaphthyl and indanylalkyldienamides gave good activity (Fig. 15), equal to that of their naphthalene analogues. The activity is very sensitive to changes in structure as is indicated by the weak activity of related analogues with differing substitution patterns and ring size.

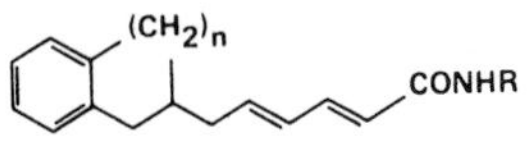

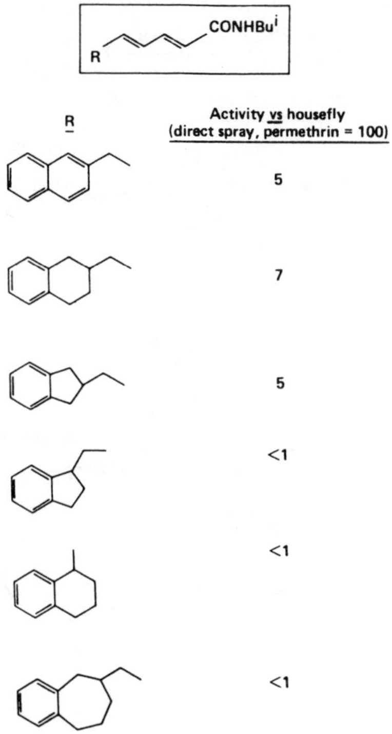

Fig. 15

The structure-activity relationships of lipid amides as a whole suggested that enhancement of potency would arise from introduction of halogenated substituents into the aromatic nucleus. For this reason suitable synthetic routes were developed. The tetrahydronaphthyl systems were prepared <u>via</u> the 1-tetralones (scheme 10), subsequent introduction of the side-chain through a malonate synthesis was followed by elaboration giving the dienamides. The analogous indanyl systems (scheme 11) were prepared from the indanones by selective allylation and reduction and ozonolysis to a precursor aldehyde.

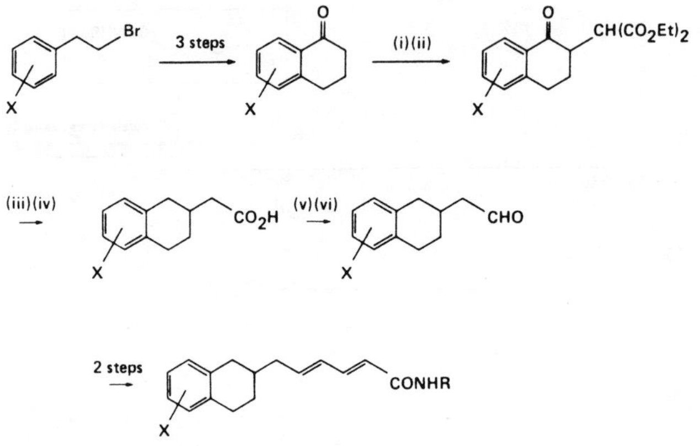

(i) Br$_2$, Et$_2$O (ii) NaCH(CO$_2$Et)$_2$, PhMe, EtOH (iii) HCl, AcOH, H$_2$O (iv) Zn/Hg, HCl, PhMe
(v) LiAlH$_4$, (vi) (COCl),–DMSO,NEt$_3$

Scheme 10

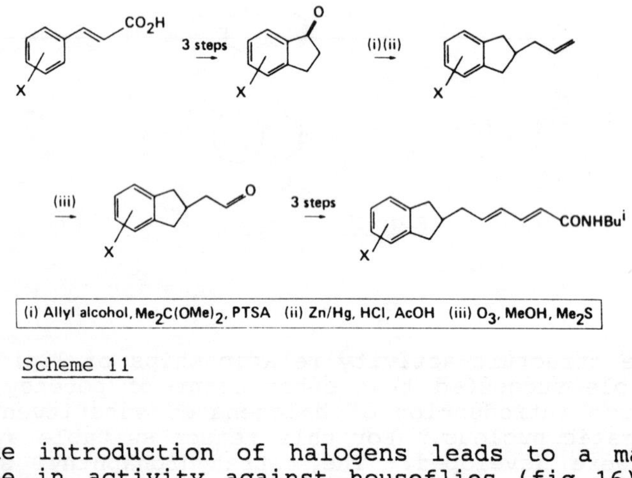

(i) Allyl alcohol, Me$_2$C(OMe)$_2$, PTSA   (ii) Zn/Hg, HCl, AcOH   (iii) O$_3$, MeOH, Me$_2$S

Scheme 11

The introduction of halogens leads to a marked increase in activity against houseflies (fig 16) with substitution by chlorine in either the C5 or C7 positions or bromine in the C5 position being advantageous. The levels of potency against mustard beetle larvae are also enhanced in this series over those of the benzyl ethers.

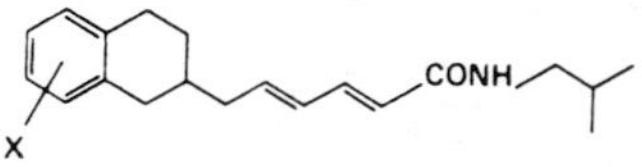

X	Housefly	Mustard beetle (larvae) permethrin = 100
H	7	2
5–Cl	23	25
5–CF$_3$	9	25
5–Br	9	
7–Cl	40	
7–Br	28	

Fig. 16

Conclusion

The lipid amides constitute a novel class of synthetic insecticidal compounds with developing levels of potency. We believe that they offer a further good illustration of the potential of natural materials as leads for the development of novel synthetic insect control agents.

Acknowledgements

The author gratefully acknowledges the contributions of the following: Co-workers at Wellcome Research Laboratories: Organic Synthesis; J.E.Robinson, S.J.Baker, D.Parkin, R.J.Peek, G.S.Cockerill; Biochemistry/ physiology; P.E.Burt, R.A.Nicholson, C.J.Brealey; Biology; M.D.V.Moss; Physical Chemistry; I.Holden, J.A.Wyatt, and M.H.Black, J.B.Weston and R.G.Wilson for their support. We are grateful to Prof.L.Crombie (Univ. of Nottingham) for much helpful discussion and with Dr. M.A.Horsham, for collaboration on novel dienamide synthesis.

References

1. M.Miyakado, I.Nakayama and N.Ohno, Insecticidal Unsaturated Isobutylamides, Insecticides of Plant Origin, Chapter 13, Am. Chem. Soc. (1989).

2. L.Crombie, J. Chem. Soc., 1955, 995

3. M.Elliott, A.W.Farnham, N.F.Janes, D.M.Johnson, D.A.Pulman and R.M.Sawicki, Agric. Biol. Chem., 50, 1347 (1986)

4. R.P.Botham, R.J.Blade, and R.A.Nicholson Pest. Sci. 1985, 555

5. G.T.Payne and D.M.Soderlund, Proc. Neurotox '88', 280 (1988)

6. C.J.Brealey, Biochem. Soc. Trans., 15, 1102 (1987)

7. R.J.Blade, R.J.Peek, J.E.Robinson and J.B.Weston, Tet. Letts., 1987, 3857

8. R.J.Blade Eur. Pat., 164, 187 (1985)

9. M.Elliott. A.W.Farnham, N.F.Janes, D.M.Johnson and D.A.Pulman. Pest. Sci., 18, 211 (1987)

10. M.Elliott et al. Pest. Sci., 18, 229 (1987)

11. R.J.Peek and R.J.Blade Eur. Pat., 228, 853 (1987)

12. R.J.Peek and R.J.Blade Eur. Pat., 209, 289 (1987)

13. D. Parkin, R.J.Blade, L.Crombie, M.A.Horsham, Eur. Pat., 228, 222 (1987)

14. P.E.Burt and R.E.Goodchild, Pest. Sci., 8, 681 (1977)

15. P.E.Burt and G.Lees, Neurotox '88', Abs. 230 (1988)

16. C.J.Brealey, Proc. Neurotox '88', 529 (1988)

17. R.J.Blade and J.E.Robinson, Tet. Letts., 1985, 3209

18. B.M.Trost, M.Lautens and B.Peterson, Tet. Letts., 1983, 4525

19. Y.Tamura, H.Maeda, H.Choi and H.Ishibashi, Synthesis, 1982, 56

20. L.Crombie, M.A.Horsham, R.J.Blade, Tet. Letts., 1987, 4879

21. M.H.Black, R.J.Blade, J.P.Larkin, J.E.Robinson and J.B.Weston, Eur. Pat., 194, 764 (1986)

New Thioureas as Insecticides

J. Drabek, M. Böger, J. Ehrenfreund, E. Stamm, A. Steinemann, A. Alder,* and U. Burckhardt.

AGRICULTURAL DIVISION AND CENTRAL RESEARCH LABORATORIES, * CIBA-GEIGY LTD., CH-4002 BASEL, SWITZERLAND

1 INTRODUCTION

Thioureas have been known to chemists since the middle of the 19th century. Since then they have been used in technical chemistry as building blocks for plastics, as vulcanization promoters in rubber chemistry and as complexing agents for heavy metals.several derivatives are active in biological systems and some found application as pharmaceuticals or animal health products[1, 2a-d]. Through research efforts within Ciba, certain thioureas were recognized as useful agents in the control of insects and mites [3], As a result, chlormethiuron (Figure 1) was developed in the early seventies into a highly successful tickicide for cattle dips.Later, Cymiazole (Figure 2), a cyclic isothiourea [4] was developed as a tickicide against multi-host ticks. In the years 1976/1977 Bayer filed patent applications (Figure 3) [5,6] claiming aromatic thioureas as acaricides and insecticides.

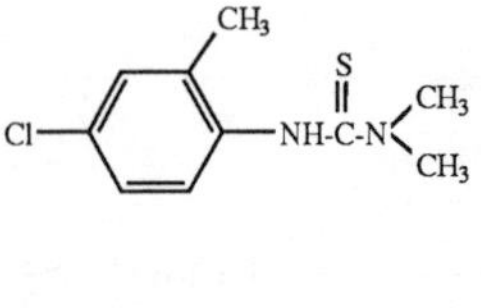

Figure 1
Chlormethiuron

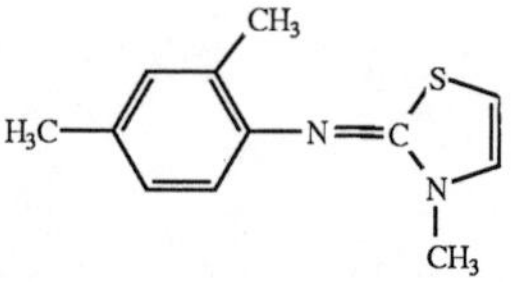

Figure 2
Cymiazole

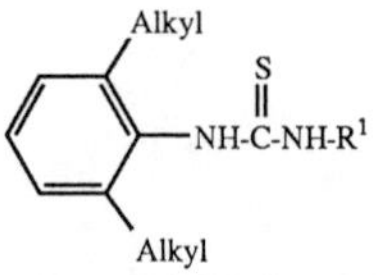

Figure 3

2 CGA 106'630, BIOLOGICAL PROFILE

We have found that the introduction of phenoxy-substituents into the para position enhances the potential for activity useful in plant protection against mites.[7] The most promising candidate of this series CGA 106'630 (Figure 4), is now being developed by Ciba-Geigy as an acaricide-insecticide for use in plant protection.

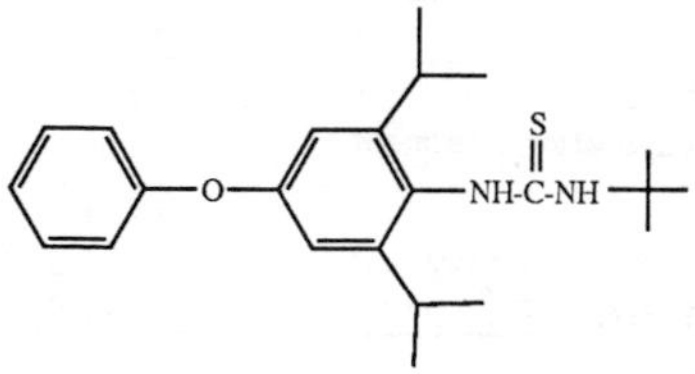

Figure 4   CGA 106'630

CGA 106'630 is highly active in laboratory and field trials against mites such as Tetranychus urticae or Tetranychus cinnabarinus but is also very effective against insects such as Bemisia tabaci, Myzus persicae or Plutella xylostella. A thorough discussion of the acaricidal and insecticidal activity of CGA 106'630 has been published recently [8].

We expect that CGA 106'630 will play an important role in the control of mites and white flies in cotton and in the control of Plutella in vegetables, especially in those areas where resistance against pyrethroids and acyl urea insecticides is already becoming apparent.

3 CHEMODYNAMIC STUDIES

It was noticeable that field trial results for CGA 106'630 often exceeded the predictions of activity that were based on the results of laboratory trials. In addition good vapour phase activity on cotton was observed despite the fact that the vapour pressure of CGA 106'630 measured in the laboratory is very low (2,2 x 10^{-7} Pa (20°C)). After a slow onset of activity excellent residual effectiveness was apparent (Figure 5). In further contrast to the laboratory observations, at higher concentrations phytotoxicity was observed in the field.

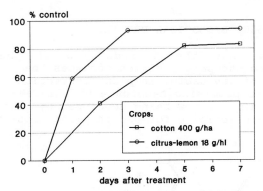

% control

<u>**Figure 5**</u> Biological activity of CGA 106'630 against
        <u>T. urticae</u> and <u>P. citri</u> in field trials.

A detailed investigation of the chemodynamic behaviour of
CGA 106'630 was carried out to resolve these
inconsistencies. The physicochemical methods and some
results used in this investigation have been recently
published [9]. While CGA 106'630 is rather stable with
regard to hydrolysis it is easily degraded
photolytically. Exposure of CGA 106'630 applied as a
droplet deposit on teflon sheets to UV-light leads to a
fast degradation. The main product of decomposition
proves to be the corresponding carbodiimide CGA 140'408
(Figure 6).

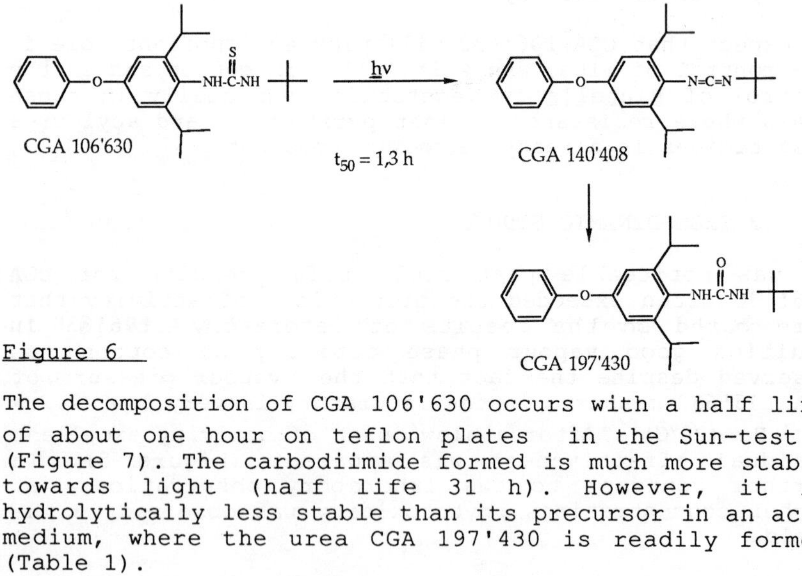

<u>**Figure 6**</u>

The decomposition of CGA 106'630 occurs with a half life
of about one hour on teflon plates in the Sun-test [9]
(Figure 7). The carbodiimide formed is much more stable
towards light (half life 31 h). However, it is
hydrolytically less stable than its precursor in an acid
medium, where the urea CGA 197'430 is readily formed
(Table 1).

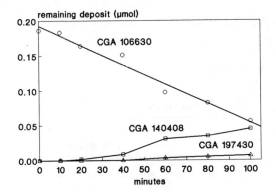

<u>Figure 7</u>   Photochemical transformation of CGA 106'630 on
teflon plates (Suntest)

<u>Table 1</u>   Hydrolysis

	Half life in hrs			Temp.
	pH 5	pH 7	pH 9	
CGA 140'408	1.1	100	>100	70°
CGA 106'630	20	21	15	95°

In order to test the photolytic conversion under field
conditions, cotton leaves were sprayed with a suspension
of CGA 106'630 and exposed to sunlight (Figure 8). Within
3 hours, half the thiourea was converted to the
carbodiimide. The concentration of CGA 140'408 never
reached the level of the starting material CGA 106'630 in
the trial on teflon plates nor in the field trials on
cotton leaves. This is in part due to the higher
volatility of the carbodiimide
(vapour pressure 9,9 x 10^{-5} Pa, 20°C).

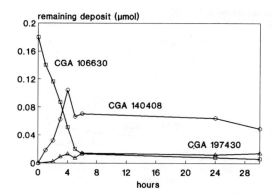

<u>**Figure 8**</u>   Photolytic transformation of CGA 106'630 on cotton leaves (SC 400 formulation field trial)

The product of slow hydrolysis, the urea CGA 197'430 emerges as the final product in the degradation sequence. In this context, it seems significant that cotton leaves have a slightly basic pH (pH 8-10) so that the hydrolytic degradation of the carbodiimide is especially slow on these leaves, and this may be the reason why CGA 106'630, via its photodecomposition product CGA 140'408, can be used successfully on cotton.

4 MECHANISM OF PHOTOLYTIC TRANSFORMATION [10]

Further studies were necessary to clarify the reaction mechanism of the photolytic transformation. Photolytic degradation experiments in solution proved that the presence of oxygen is essential for the progress of the reaction (Figure 9). Furthermore the solution of the reaction turned acidic due to the sulfuric acid formed in the process.

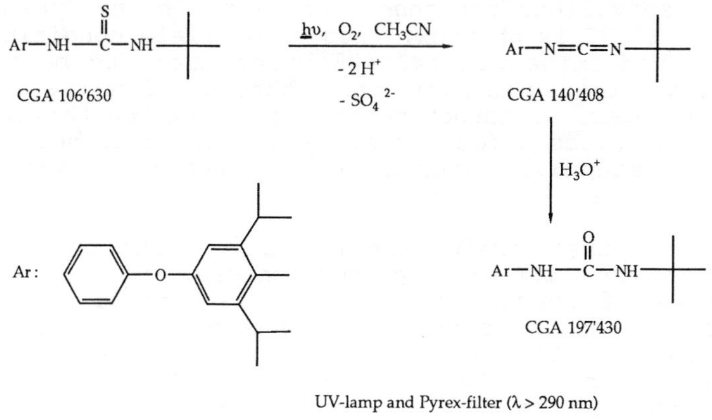

UV-lamp and Pyrex-filter ($\lambda > 290$ nm)

Figure 9

From this evidence, we suspected that singlet oxygen
might be the active species in this photo-oxidation.
Singlet oxygen can be formed by the reaction of triplet
oxygen either with a photochemically excited state of the
substrate or with an externally added sensitizer. Indeed,
addition of sensitizers like rose Bengal or Methylene
blue enhanced the rate of photolytic degradations
dramatically. Quencher experiments provided further
evidence for the involvement of singlet oxygen. Addition
of singlet oxygen quenchers such as NaN3, DABCO or
2,2,6,6-TMP slowed the reaction down significantly.

5 BIOLOGY OF TRANSFORMATION PRODUCTS

The two main degradation products from the photooxidation
of CGA 106'630 display very different biological
activities. The carbodiimide 140'408 is superior in its
activity against mites, chewing insects and sucking
insects.

It is however, in contrast to the precursor, phytotoxic
towards a number of important crops (rice, cotton, soya,
beans, cabbage). The urea CGA 197'430, on the other hand
is essentially devoid of any biological activity.

All our observations are consistent with the hypothesis that CGA 106'630 is photooxidized under field conditions to the carbodiimide CGA 140'408 which seems to be the true active ingredient in the control of mites and insects. However, it cannot be excluded that the thiourea also has a direct effect on mites, since we do not yet know the metabolic paths of either acaricide within mites or insects.

It might be asked why we are not developing the true active ingredient for plant protection use. The answer is that the use of the thiourea as a proinsecticide instead of the corresponding carbodiimide has many advantages. The thiourea CGA 106'630 seems to act as a slow-release formulation for the active ingredient. The important vapour phase activity of the carbodiimide CGA 140'408 can therefore be maintained for a much longer time in the field. Furthermore possible phytotoxicity problems can be better coped with due to the small initial concentration of the carbodiimide. In chemical synthesis the thiourea serves as a logical precursor for the carbodiimide: we can therefore save the last synthetic step.

6 THIOUREA DERIVATIVES

Independently from these physicochemical studies derivatives of thioureas were synthesized mainly for chemical reasons. It was recognized early on that many isothioureas and carbodiimides derived from the active thioureas display interesting insecticidal and acaricidal activities. An example is shown in Figure 10.

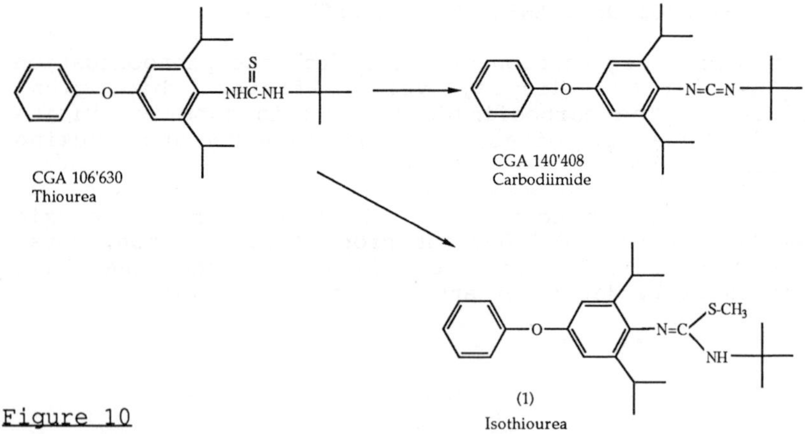

CGA 106'630
Thiourea

CGA 140'408
Carbodiimide

(1)
Isothiourea

Figure 10

From the many examples of carbodiimides synthesized, the pyridine-derivative (2) displayed especially high acaricidal activities in citrus crops (Figure 11).

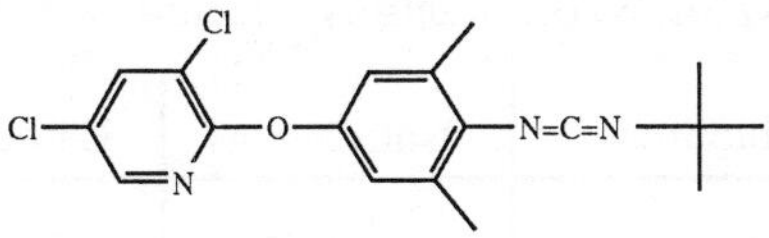

(2)

Figure 11

The activities in the laboratory of the thiourea, isothiourea and the two carbodiimides against insects and mites are compared in the following table (Table 2).

Table 2 Activity against insects and mites

Minimal conc. in ppm for > 80% control

CGA-No.	Spodoptera littoralis	Nilaparvata lugens	Myzus persicae	Tetranychus urticae
106'630	50	200	400	26
140'408	25	100	12	6
(1)	25	200	200	100
(2)	200	>400	>400	6

The three classes of new insecticides/acaricides thioureas, isothioureas and carbodiimides have different potency profiles in their biological activities. The spectrum of biological activities is summarized in Table 3. The importance of different side effects also varies within the three classes of compounds as shown in Table 4.

Table 3   Biological activity

Activity against	$\overset{\overset{\text{S}}{\|\|}}{\text{-NH-C-NH-}}$ Thioureas	$\text{-N=C}\overset{\text{S-CH}_3}{\underset{\text{NH-}}{\diagup}}$ Isothioureas	-N=C=N- Carbodiimides
Lepidoptera	++	+++	+++
Plant hoppers	+	+++	++
Aphids	+	++	++
Spider mites	++	+	+++

Table 4   Side effects

Side effects	$\overset{\overset{\text{S}}{\|\|}}{\text{-NH-C-NH-}}$ Thioureas	$\text{-N=C}\overset{\text{S-CH}_3}{\underset{\text{NH-}}{\diagup}}$ Isothioureas	-N=C=N- Carbodiimides
Acute toxicity rats	low	high	low
Fish toxicity	variable	variable	high
Phytotoxicity	low	low	high

From the last two tables, it is apparent that isothioureas might be especially interesting candidates if the inherent problems of mammalian toxicity could be eased. It is well known from the literature [11] that the concept of proinsecticides can be successfully applied for this purpose. Indeed, if isothioureas are transformed by acylation into the corresponding oxalic acid amides, a dramatic detoxifying effect can be observed (Table 5).

Table 5 Acylated Isothioureas
Toxicity

LD_{50} rat
p.o.   mg/kg

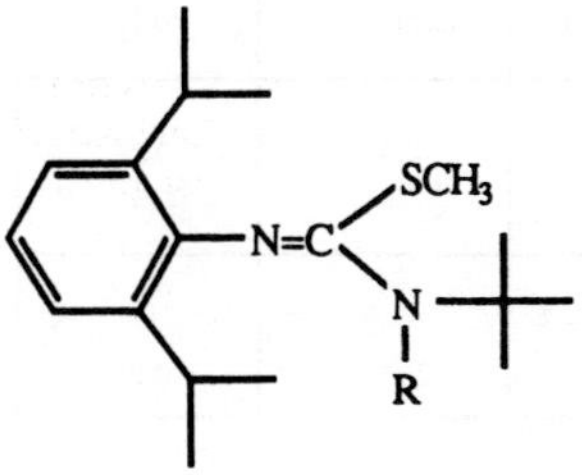

R = H   10-50

R = COCOO—$\left<\right.$ > 200

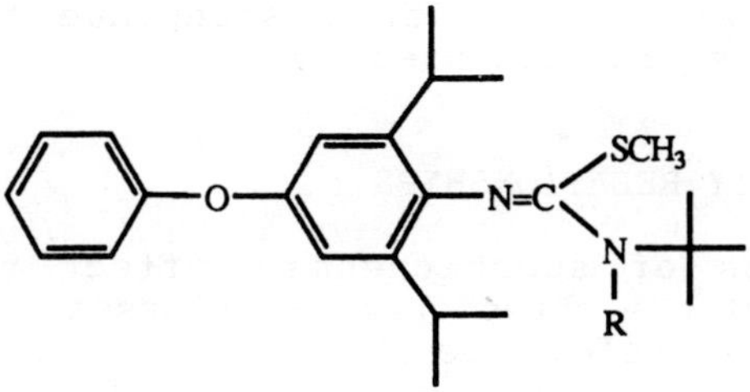

R = H   8

R = COCOO—$\left<\right.$ 478

In general the high insecticidal activity of the
Isothiourea moiety is maintained in the proinsecticides
(Table 6).

<u>Table 6</u> Acylated Isothioureas
Activity

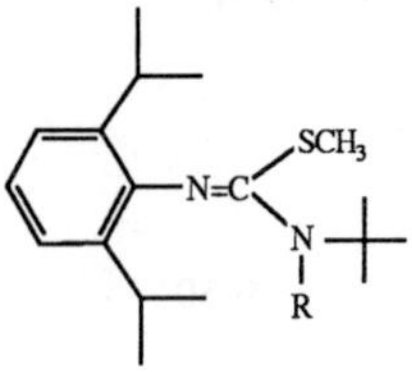
SCH$_3$
N=C
N
R

	Minimal concentration for 80% activity in ppm			
	Nilaparvata lugens		Nephotettix sp.	Diabrotica sp.
	systemic	contact	contact	soil
R = H	0.75	100	400	0.75
R =COCOO	0.75	400	400	0.75

Unfortunately the oxalic acid amides proved to be too labile with regard to hydrolysis, especially in formulations. Since all efforts to stabilize the formulation have failed, we had to discontinue the development of these promising proinsecticides.

7 STRUCTURE-ACTIVITY RELATIONSHIPS :

In general, variations of substituents affect the biological activities of the three chemical classes in a similar way and will therefore be discussed globally (Figure 12).

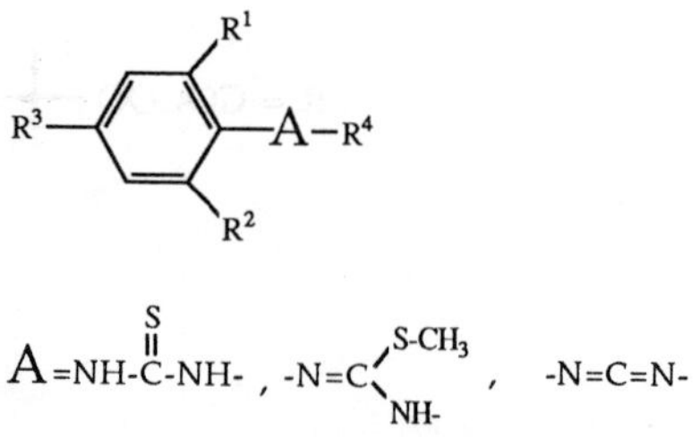
R^1
R^3
A–R^4
R^2
A =NH-C(S)-NH- , -N=C(S-CH$_3$)(NH-) , -N=C=N-

R_1, R_2 = Alkyl, Cycloalkyl
R_3 = X-Aryl, X-Heterocyclyl, X = O, S, CH_2
R_4 = sec. oder tert. Alkyl.

<u>Figure 12</u>

For good activities, the <u>ortho</u> positions must be occupied by alkyl- or cycloalkyl groups. Optimally, these groups should be both isopropyl or one ethyl and one isopropyl. Replacements by hydrogen, halogen, trifluromethyl- or methoxy-groups lead to decreased biological activity.

The character of the substituent R3 partially determines the spectrum of biological activities. Short substituents in the <u>para</u>-position (hydrogen, alkyl, <u>S</u>-alkyl) lead to compounds active against rice and soil insects, whereas aryl-substituents connected via a bridge give rise to representatives active against lepidoptera and particularly mites. R4 has to be a secondary or tertiary alkyl substituent for good activity, the optimum being the tertiary butyl group. Straight chain aliphatic or phenyl groups decrease the insecticidal activity.

8 SYNTHESIS OF THIOUREAS AND CARBODIIMIDES

The anilines can easily be converted into the isothiocyanates either by reaction with thiophosgene in basic media or, on a large scale preferably by reaction with sodium rhodanide and pyrolysis of the thiourea formed. Tertiary butylamine is added to the isothiocyanate to give thioureas such as CGA 106'630.(Figure 13)

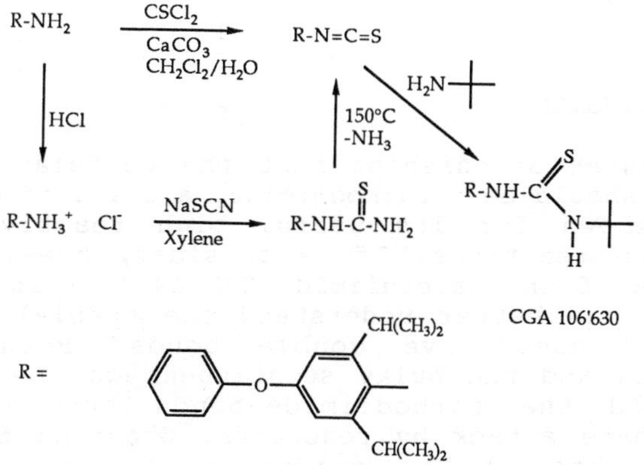

<u>Figure 13</u>

The carbodiimides can be prepared handily in the laboratory by Mukaiyama's method [12] (Figure 14). Another pathway, more promising for large scale preparation, was developed from our knowledge about the reaction mechanism of the photolytic degradation of thioureas such as CGA 106'630. The thiourea can be converted with very high yield to the carbodiimide CGA 140'408 by preparative photooxydation using a sodium-lamp as a light source and rose Bengal or Methylene blue as a sensitizer [13].

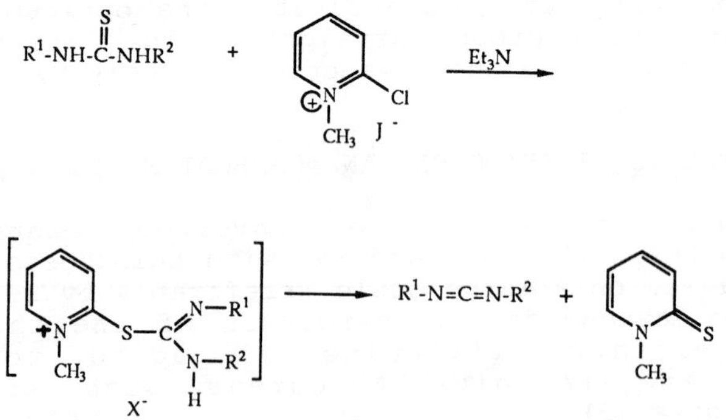

$$R^1\text{-NH-}\overset{\overset{\displaystyle S}{\|}}{C}\text{-NHR}^2 \quad + \quad \text{(pyridine ring)} \quad \xrightarrow{\text{Et}_3\text{N}}$$

$$\left[\text{(structure)} \right] \quad \longrightarrow \quad R^1\text{-N=C=N-R}^2 \quad + \quad \text{(pyridinethione)}$$

<u>Figure 14</u>

9 CONCLUSION

It is somewhat astonishing that the postulated active ingredient should be a carbodiimide, a class of compounds normally known for its useful high reactivity in condensation reactions. If we consider, however, the exact shape of the carbodiimide CGA 140'408 in computer models, we can better understand the special geometry around the cumulative double bonds. Both <u>ortho</u>-substituents and the bulky substituent on the nitrogen atom shield the carbodiimide-bonds from easy and indiscriminate attack by reactants. Choosing the right size for these bulky groups seems to guarantee selectivity in biological performance and the necessary hydrolytic stability in the formulations.

With the development of CGA 106'630 we hope to give the farmer a new and useful tool to solve his problems with sucking insects and mites - be it finally the action of a thiourea or a carbodiimide.

REFERENCES

1. Ullmanns Enzyklopädie der technischen Chemie, 4th edition, Verlag Chemie 1983, Vol 23, p. 172.
2. M. Negwer, Organic-chemical drugs and their synonyms, 6th edition. Akademie Verlag Berlin 1987. a) 728, 1169, b) 715, 134, 2635, c) 6494, 6971, d) 3213, 5406, 3613.
3. D. Dürr and V. Dittrich, DE 1542848.
4. D. Dürr and W. Traber, DE 2619724.
5. a) E. Enders, W. Stendel and I. Hammann, DE 2639748 b) E. Enders, W. Stendel and I. Hammann, DE 2657772 b) E. Enders, I. Hammann and W. Stendel, DE 2702235.
6. E. Enders, I. Hammann and W. Stendel, DE 2727529.
7. J. Drabek and M. Böger, DE 3034905.
8. H.P. Streibert, J. Drabek and A. Rindlisbacher. Proceedings Brighton Crop Protection Conference - Pests and Diseases, 1988, 25.
9. A. Steinemann, E. Stamm and B. Frei, Aspects of Applied Biology, 1989, 21, 203-213.
10. A. Alder et al., Paper in press.
11. J. Drabek, R. Neumann "Proinsecticides" in "Insecticides". D.H. Hutson, T.R. Roberts Eds. John Wiley, 1985, p. 35-86.
12. T. Shibanuma, M. Shiono, T. Mukaiyama, Chem. Lett. 1977, 575.
13. A. Alder, EP 307361.

The Synthesis and Properties of N-Sulphenyl Acylureas

M. Anderson, * A. G. Brinnand, P. Camilleri, †E. J. Langner, and R. C. Weaver††

SHELL RESEARCH LTD., SITTINGBOURNE RESEARCH CENTRE, SITTINGBOURNE, KENT ME9 8AG, UK

The field performance of an insecticide is strongly dependent upon how effectively it can be distributed within a crop. Such materials are usually applied as finely-divided suspensions or emulsions produced by diluting a concentrated liquid formulation with water. For non-volatile solids the most efficient coverage is obtained by applying them in solution. This is, of course, only possible if the physical properties of the active material are suitable and in many cases they are not.

The acylurea insecticides (also known as benzoylphenylureas, Fig 1) are characteristically high-melting solids with low solubility in most solvents. Such physical properties make the preparation of liquid formulations very difficult and earlier compounds such as diflubenzuron were usually applied as wettable powders. The rate of absorption of such insoluble materials into a target pest (and hence their performance) is strongly dependent upon particle size, the finer suspensions having the highest activity. In the case of diflubenzuron this led to the development of wettable powder formulations in which the average particle size is typically less than 5 microns[1].

+ Current address: Smith Kline and French Research Ltd., The Frythe, Welwyn, Herts AL6 9AR, UK.

++ Current address: Pfizer Ltd., Ramsgate Road, Sandwich, Kent CT13 9NJ

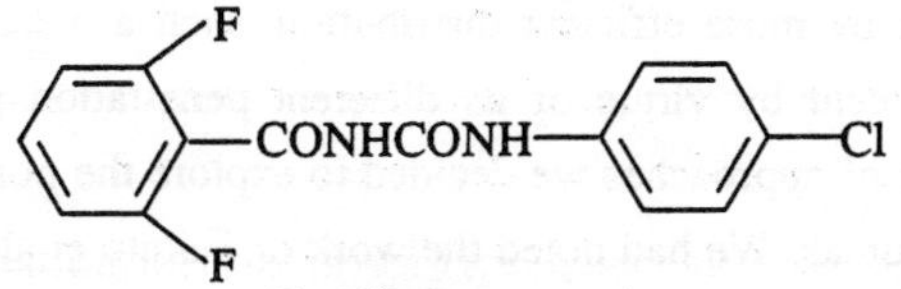

(1, diflubenzuron)

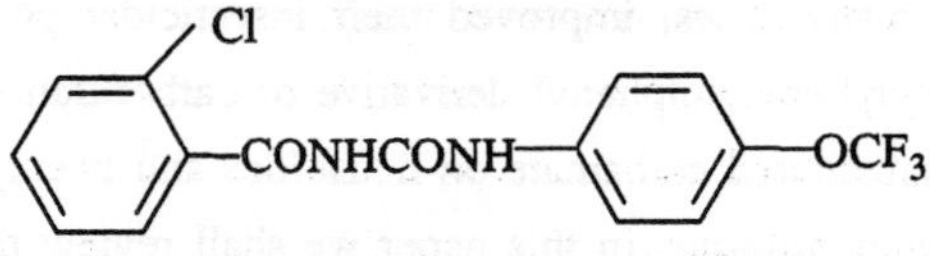

(2, triflumuron)

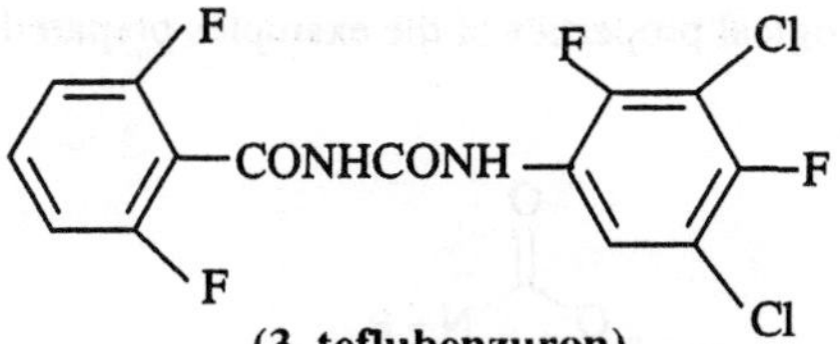

(3, teflubenzuron)

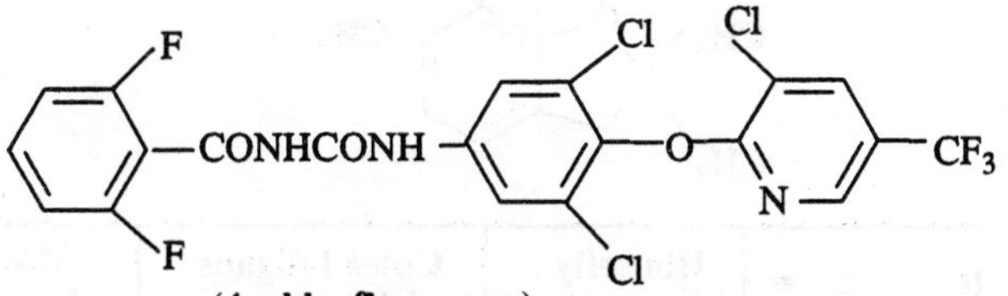

(4, chlorfluazuron)

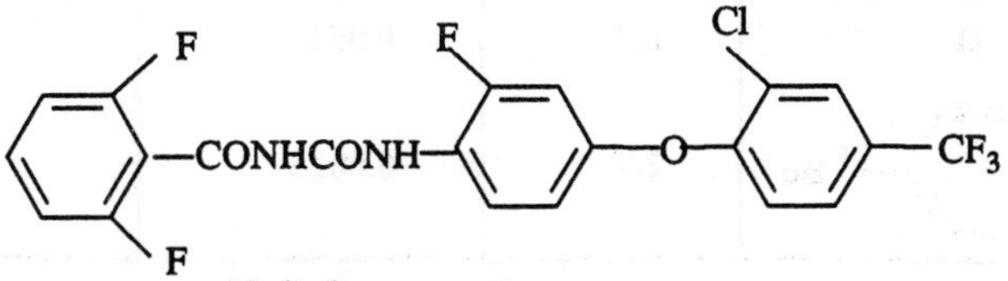

(5, flufenoxuron)

Figure 1: Structures of some typical acylurea insecticides

Upon entering this field of research in 1980 we reasoned that further improvements in performance might be possible if the limitations imposed by physical properties on formulation could be overcome. One of our first objectives was to discover an acylurea insecticide with sufficient solubility in organic solvents to facilitate formulation as an emulsifiable concentrate. Apart from any benefits in performance brought about by more efficient distribution, such a material might also be inherently more potent by virtue of its different penetration properties. After considering a number of approaches we decided to explore the possibility of preparing $\underline{N}$-sulphenyl acylureas. We had noted the work of Fukuto et al[2] who had shown that $\underline{N}$-sulphenylation of carbamate insecticides not only increased their lipophilicity but also, in many cases, improved their insecticidal potency. For example, the $\underline{N}$-(4-*tert*-butylphenyl)sulphenyl derivative of carbofuran (Fig. 2) is twice as active as the unsubstituted carbamate on houseflies and twenty times as active on the mosquito *Culex fatigans*. In this paper we shall review the various synthetic approaches we took towards $\underline{N}$-sulphenyl acylureas and describe some of the chemical, physical and biological properties of the examples prepared.

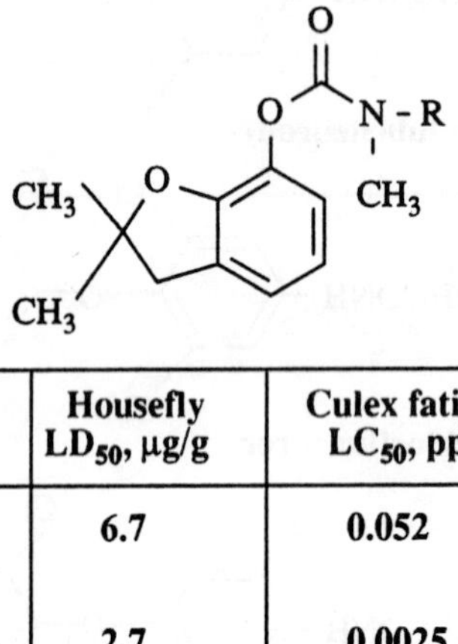

R	Housefly LD_{50}, µg/g	Culex fatigans LC_{50}, ppm	Mouse oral LD_{50}, mg/kg
H	6.7	0.052	2
 $-S-\langle\text{ring}\rangle-^{t}Bu$	2.7	0.0025	75

Figure 2: Effect of $\underline{N}$-sulphenylation on the toxicity of carbofuran (Ref. 2)

1.   Synthetic approach

Much of the exploratory synthesis was carried out with well-characterised acylurea insecticides so that any influence of sulphenylation on biological performance could be more easily monitored.

1.1   Direct N-sulphenylation

The usual technique for preparing N-sulphenyl derivatives of methylcarbamate insecticides is to react the appropriate sulphenyl halide with the carbamate in the presence of a tertiary organic base such as pyridine or triethylamine:

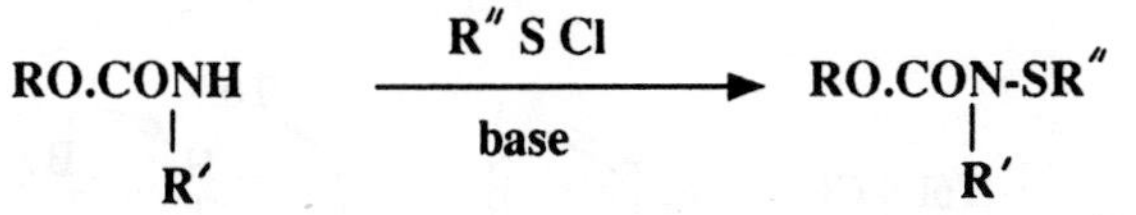

$$RO.CONH \quad \xrightarrow[\text{base}]{R''SCl} \quad RO.CON\text{-}SR''$$
$$| \qquad\qquad\qquad\qquad |$$
$$R' \qquad\qquad\qquad\qquad R'$$

When an attempt was made to react triflumuron (2) with *p*-toluenesulphenyl chloride in pyridine little change occurred, but in DMF containing an equivalent of triethylamine, reaction was complete within $3\frac{1}{2}$ h at room temperature. Two major products were formed and these were identified as 2-chloro-N,N-bis(*p*-tolylthio)benzamide (7) and 2-chloro-N-(4-trifluoromethoxy)benzamide (8) (Scheme 1). Nothing resembling any of the anticipated N-sulphenyl products (6) could be found in the reaction mixture. Evidently an extremely facile fragmentation had occurred.

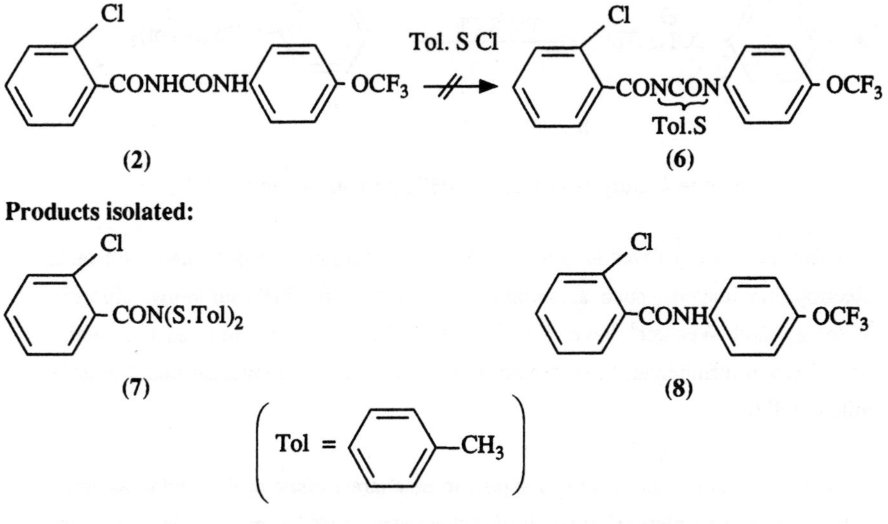

Products isolated:

$$\left(\text{Tol} = \text{⟨} \quad \text{⟩}\text{-CH}_3 \right)$$

Scheme 1: Attempted sulphenylation of triflumuron (2)

Possible mechanisms for the formation of (7) and (8) are outlined in Schemes 2 and 3. Both invoke the $\underline{N}$-sulphenyl derivative (9) as the primary product, which is then transformed further by two different base-catalysed elimination reactions. The formation of (8) at first seems rather surprising, but is clearly analogous to the formation of amides by the decarboxylation of mixed carboxylic/carbamic anhydrides[3] (Scheme 3).

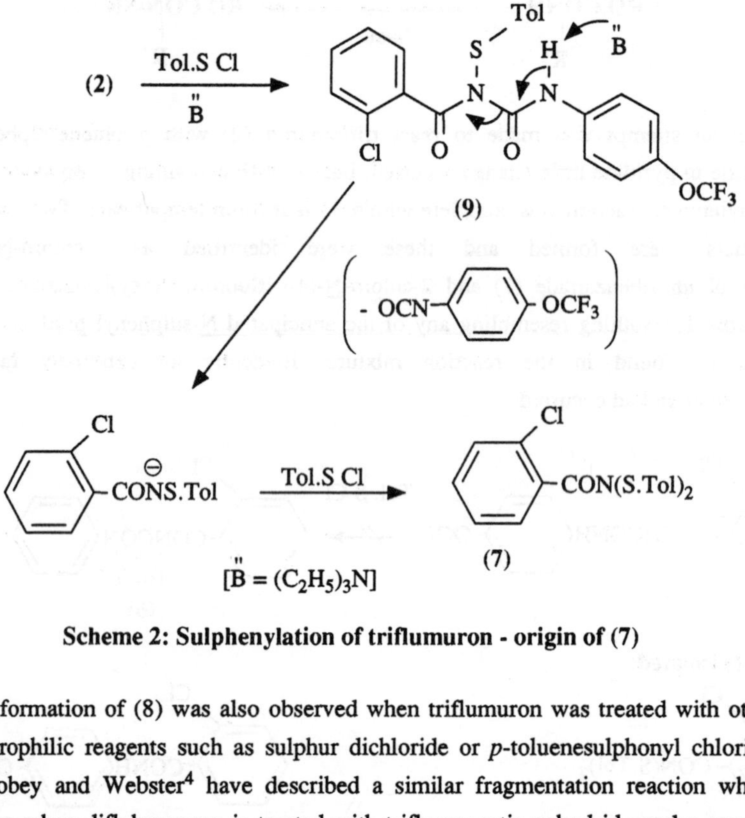

Scheme 2: Sulphenylation of triflumuron - origin of (7)

The formation of (8) was also observed when triflumuron was treated with other electrophilic reagents such as sulphur dichloride or *p*-toluenesulphonyl chloride. Worobey and Webster[4] have described a similar fragmentation reaction which occurs when diflubenzuron is treated with trifluoroacetic anhydride under equally mild conditions.

These facile elimination reactions appear to be characteristic of the acylureas and it is tempting to speculate whether similar chemistry could be involved in their mode of action.

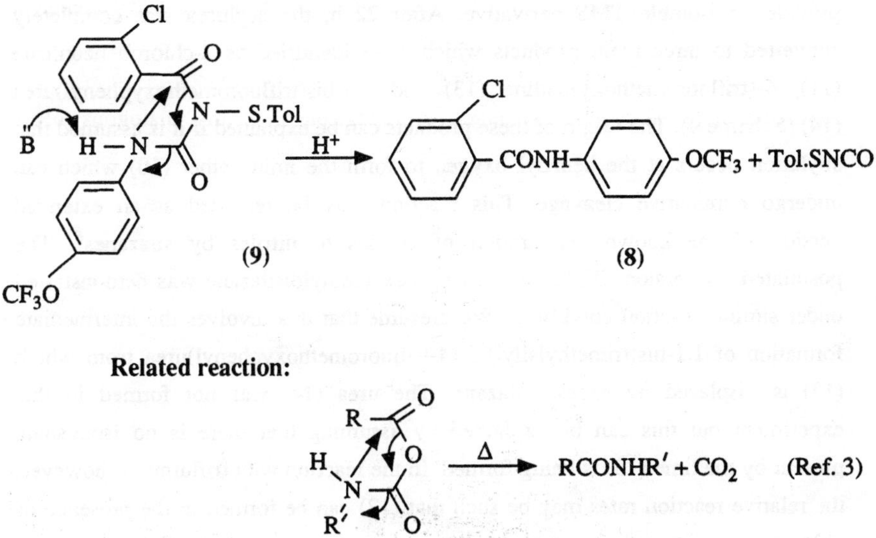

(9)   **(8)**

Related reaction:

$$\xrightarrow{\Delta} RCONHR' + CO_2 \quad (Ref. 3)$$

Scheme 3: Sulphenylation of triflumuron - origin of (8)

1.2   Sulphenylation *via* N-silyl intermediates

Having failed to introduce a sulphenyl group by the direct method, it was decided to adapt a technique developed by Harpp[5] for the sulphenylation of cyclic imides which utilises N-trimethylsilyl intermediates:

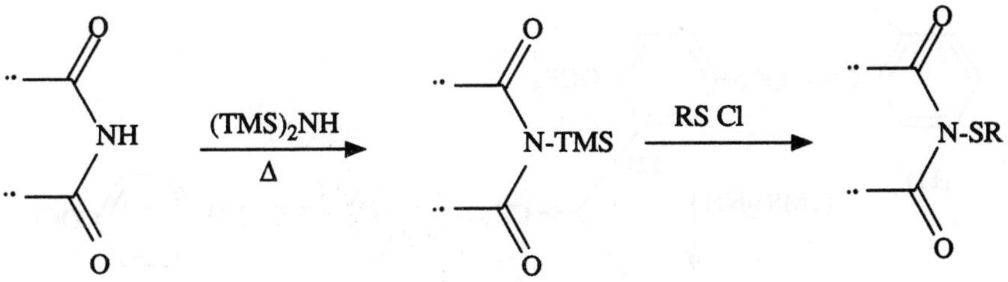

$$[TMS = Si\,(CH_3)_3]$$

Treatment of triflumuron with refluxing hexamethyldisilazane containing a catalytic amount of imidazole, according to the published procedure, failed to provide an isolable TMS derivative. After 22 h, the acylurea was completely converted to three main products which were identified as 2-chlorobenzonitrile (11), 4-(trifluoromethoxy)aniline (13) and 1,3-bis(trifluoromethoxyphenyl)urea (14) (Scheme 4). The origin of these products can be explained if it is assumed that silylation occurs at the benzoyl oxygen to form the imino ether (10) which can undergo eliminative cleavage. This reaction may be regarded as an extended version of the known dehydration of amides to nitriles by silazanes[6]. The postulated conversion of (12) to (13) by hexamethyldisilazane was demonstrated under similar reaction conditions. We presume that this involves the intermediate formation of 1,1-bis(trimethylsilyl)-3-(4-trifluoromethoxyphenyl)urea from which (13) is displaced by excess silazane. The urea (14) was not formed in this experiment but this can be explained by assuming that there is no isocyanate present by the time (13) is being formed. In the reaction with triflumuron however, the relative reaction rates may be such that (12) can be formed in the presence of (13).

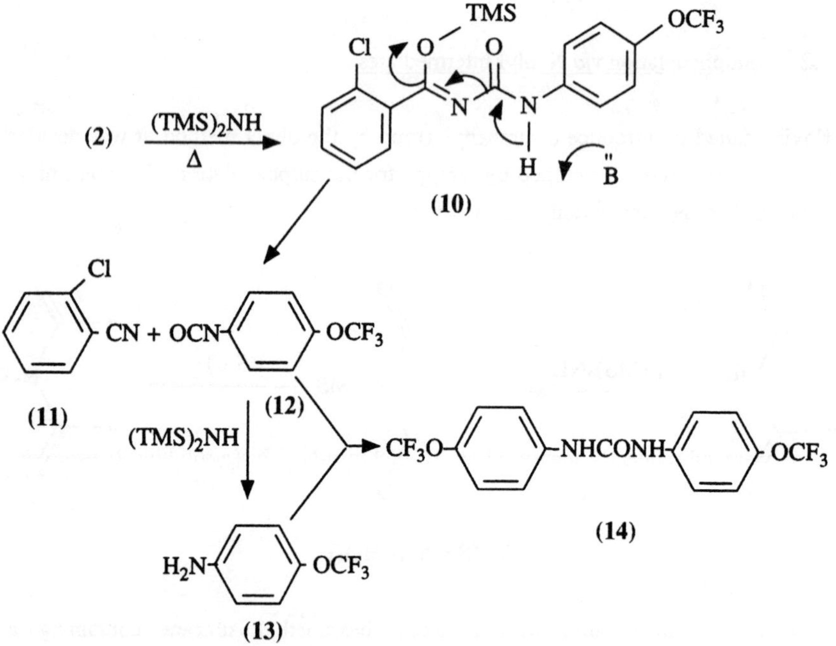

Scheme 4: Silylation of triflumuron (2)

1.3 The convergent approach

In view of the foregoing, it was decided to examine a convergent synthetic approach, i.e. introduction of the sulphenyl substituent *via* a sulphenylated intermediate.

(a) Ar CONHR + Ar'NCO ⟶ ArCONCONHAr'
 (15)   (16) |
                               R   (17)

(b) ArCONCO + Ar'NHR ⟶ ArCONHCONAr'
 (18)   (19)   (20) |
                                     R

Scheme 5: Synthesis routes to acylureas

Referring to Scheme 5, the two most common procedures for the synthesis of acylureas are (a), the reaction of an amide with an isocyanate or (b), the reaction of an acyl isocyanate with an amine. By using these methods it is theoretically possible to introduce a substituent R specifically onto either of the acylurea nitrogen atoms. Amides of the type (15; R=SR') can be obtained in good yield by sulphenylating the corresponding N-TMS intermediates[7] and the sulphenamides (19; R=SR') can be prepared by sulphenylation of the appropriate anilines under the conditions described below.

Applying method (a), 2,6-dichloro-N-(*p*-tolylthio)benzamide and 4-chlorophenyl isocyanate failed to react when heated together in refluxing toluene. When the same reactants were dissolved in toluene containing a trace of triethylamine however, a rapid reaction took place at room temperature to give two products. These were subsequently identified as 2,6-dichloro-N,N-bis(*p*-tolylthio)benzamide (23) and 1-(4-chlorophenyl)-3-[2,6-(dichlorobenzoyl)]urea (24) (Scheme 6). A plausible explanation of this result is that the sulphenylated acylurea is formed initially, but this reacts with more of the sulphenyl amide (21) to form the observed products. (N-sulphenylated imides can be used as sulphenylating agents[8]).

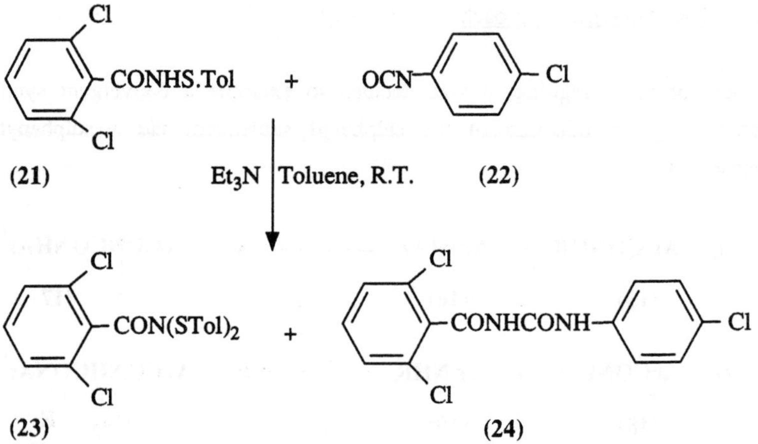

Scheme 6: Attempted synthesis of an acylurea sulphenylated
on the imide nitrogen (method a)

The alternative convergent approach (method b) proved to be more successful and has subsequently been used to prepare over 300 novel sulphenyl acylureas in this laboratory. The sulphenamides (19; R=SR′) were obtained by the reaction of sulphenyl halides with the appropriate aniline in the presence of excess base. If pure sulphenyl halide is available, a solution of the halide in ether or methylene chloride is added to an equivalent amount of the aniline in the same solvent containing a slight excess of triethylamine[9]. Reaction is complete within 2-4 hours at 5-10° and the yield of sulphenamide is usually 60-90%. In cases where the required sulphenyl halides are not readily isolable, it is more convenient to prepare them *in situ* by halogenation of the corresponding thiol or disulphide[7] in methylene chloride and to use the resulting solution to sulphenylate the aniline. Yields obtained in this way are comparable with those obtained by the direct method. By using these techniques a range of S-arylsulphenamides were prepared which were usually isolable as stable crystalline solids. With the exception of the trichloromethanesulphenyl derivatives, S-alkylsulphenamides were too unstable to isolate.

Condensation of the sulphenamides with aroyl isocyanates is preferably carried out at room temperature in an aprotic solvent. Hydrocarbons, chlorinated hydrocarbons and ethers are all suitable and it is usually possible to select a solvent system from which the product will crystallise preferentially during reaction. Highly soluble sulphenyl acylureas were isolated by chromatography on silica gel. Given that the sulphenamides (19) can be prepared, route (b) appears to have few limitations. The only failures we have encountered are in cases where the sulphenyl moiety contains a pyridyl group.

2.   The properties of N-sulphenyl acylureas

Having established a general synthetic method, the next task was to identify structural features which would provide a material with high biological activity and sufficient stability and solubility to be applied as an emulsifiable concentrate. Hydrolytic stability of the N-S linkage is particularly important if any benefit in performance is to be gained from a change in physical properties brought about by sulphenylation. It is also important with regard to bioassays where any breakdown during the test (in caterpillar diet, for example) would lead to erroneous results.

A study of the hydrolysis of a series of substituted N-phenylsulphenyl derivatives of triflumuron (Table 1) showed that cleavage of the N-S bond is base-catalysed and that for a given pH, the rate can vary considerably with substitution. Thus, electron-withdrawing substituents on the phenylthio group increase the rate of hydrolysis (compare the 4-nitrophenyl with the 4-tolyl compound). The rate of hydrolysis is reduced by substitution *ortho-* to the sulphur atom, presumably due to steric shielding. The exceptional stability of the 2-nitro compounds however, appears to be another example of the specific *ortho*-effect observed with nitrophenylsulphenyl derivatives[10] (see later). It was concluded at this stage that 2-nitrobenzenesulphenyl derivatives would provide the most reliable screening results and that these were the only groups with sufficient stability to be considered for further examination.

Table 1: Hydrolysis of the N-S bond in some N-sulphenyl acylureas

| R | Half-life (h) | | | Temp |
	pH 5[a]	pH 7	pH 9	°C
4-NO$_2$	0.78	0.06[b]	c	20
4-Me	4.0	0.17	c	20
2-Me	21.0	0.66	c	20
2-^{i}Pr	37.0[b]	1.25	0.10	20
2-NO$_2$	2625.0	176.0	65.0	30
2,4-(NO$_2$)$_2$	924.0	54.0	37.0[b]	30

a: The pH value quoted is that of the aqueous buffer.
b: Results obtained by extrapolation.
c: Too rapid to measure.

The effect of the introduction of the N-(2-nitrobenzenesulphenyl) substituent on insecticidal activity is illustrated in Table 2. The columns labelled S.l., S.l.ov. and A.a. refer to activity against the Egyptian cotton leaf worm (*Spodoptera littoralis*) the eggs of *S. littoralis* and larvae of the yellow fever mosquito (*Aedes aegypti*) respectively. The *Spodoptera* larvae (2nd instar) were fed on treated diet and mortality was assessed after seven days. The ovicidal activity was determined by spraying freshly laid eggs with solutions of the test compound and counting those which failed to hatch after five days. The activity against the mosquito was determined by placing 4th instar larvae in a fine aqueous suspension of the test compound for 48 hrs. They were then supplied with nutrient and mortality was assessed by measuring the proportion successfully emerging as adults (usually after about 10 days).

Table 2: Insecticidal activity of 2-nitrobenzenesulphenyl derivatives

Compound	R_1	R_2	R_3	Toxicity Index[a]		
				S.l.	S.l.ov.	A.a.
25	F	F	Cl	90 *100*	270 *5100*	220 *73*[b]
26	F	F	CF_3	160 *290*	1800 *6100*	180 *77*
27	Cl	H	OCF_3	750 *430*	210 *5900*	220 *130*
28	F	F		4500 *6700*	c *c*	310 *780*

a: Toxicity index = $\dfrac{LC_{50}\ \text{Parathion}}{LC_{50}\ \text{Compound}} \times 100$

b: The numbers in italics refer to the activity of the corresponding unsubstituted acylurea.

c: Inactive at the highest dose used (0.2%).

Compounds (25), (26) and (27) are related to the acylureas diflubenzuron, penfluron and triflumuron which are characterised by larvicidal and ovicidal activity on *Spodoptera*. Compound (28) is representative of the aryloxyphenyl ("second generation") acylureas which, in our screen, are potent *Spodoptera* larvicides, but rarely ovicidal. As shown in Table 2, the spectrum of activity of the sulphenyl derivatives parallels that of the corresponding unsubstituted compounds and in some cases there is an improvement in potency. See, for example, compound (27) on *Spodoptera* and compounds (25), (26) and (27) on *Aedes*. Although the introduction of an N-(2-nitrobenzenesulphenyl) substituent slightly improves solubility (Table 3) this is not sufficient to enable formulation as an emulsifiable concentrate. When compounds (27) and (28) were evaluated in the field as wettable powders however, we were encouraged to find that they performed as well or slightly better than the corresponding unsubstituted acylureas. Our next objective was to discover an $\underline{N}$-sulphenyl substituent with similar hydrolytic stability to the 2-nitrobenzenesulphenyl group which would have a greater impact on solubility.

Table 3: The effect of the $\underline{N}$-(2-nitrobenzenesulphenyl) substituent on xylene solubility

Compound	R_1	R_2	R_3	R_4	Xylene sol. g/l
2	Cl	H	OCF$_3$	H	2.3
27	Cl	H	OCF$_3$	$-S-$ (2-nitrophenyl)	3.7
29	F	F	(Cl, O, CF$_3$ phenyl)	H	5.7
28	F	F	(Cl, O, CF$_3$ phenyl)	$-S-$ (2-nitrophenyl)	18.8

The exceptional hydrolytic stability of 2-nitrobenzenesulphenyl derivatives has been attributed to a specific non-bonded interaction between the sulphur atom and one of the oxygen atoms on the nitro group. In the crystal structure of methyl 2-nitrobenzenesulphenate for example, the nitro group and benzene ring are coplanar and the distance between the sulphur atom and the neighbouring nitro oxygen atom is considerably shorter than the sum of their van der Waals radii[10]. With the nitro group fixed in this configuration the alignment of the electron-rich oxygen atom is such that it offers both a steric and electrostatic impediment to nucleophilic attack at sulphur. A similar situation can be envisaged with the corresponding acylurea derivatives (Fig. 3a).

Figure 3

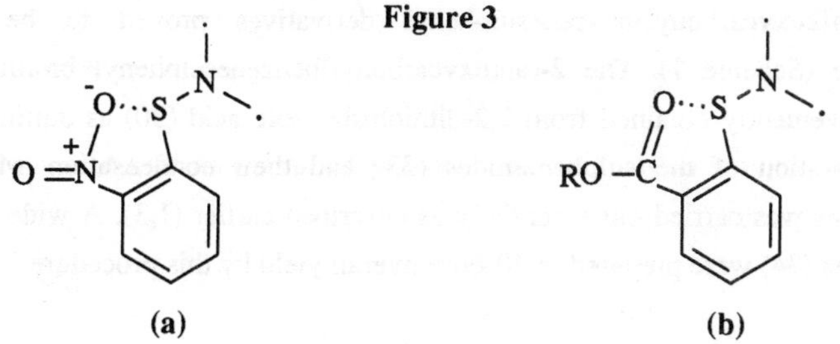

(a)   (b)

It has been postulated that the stability of some anthroquinone sulphenyl derivatives could be due to a similar interaction between the sulphur atom and a neighbouring carbonyl group[10]. With this latter point in mind, we decided to examine the 2-(alkoxycarbonyl)benzenesulphenyl group as an N-substituent (i.e. Fig. 3b). Apart from any stabilisation afforded by the carbonyl group, the ester function would provide a versatile means for the manipulation of physical properties.

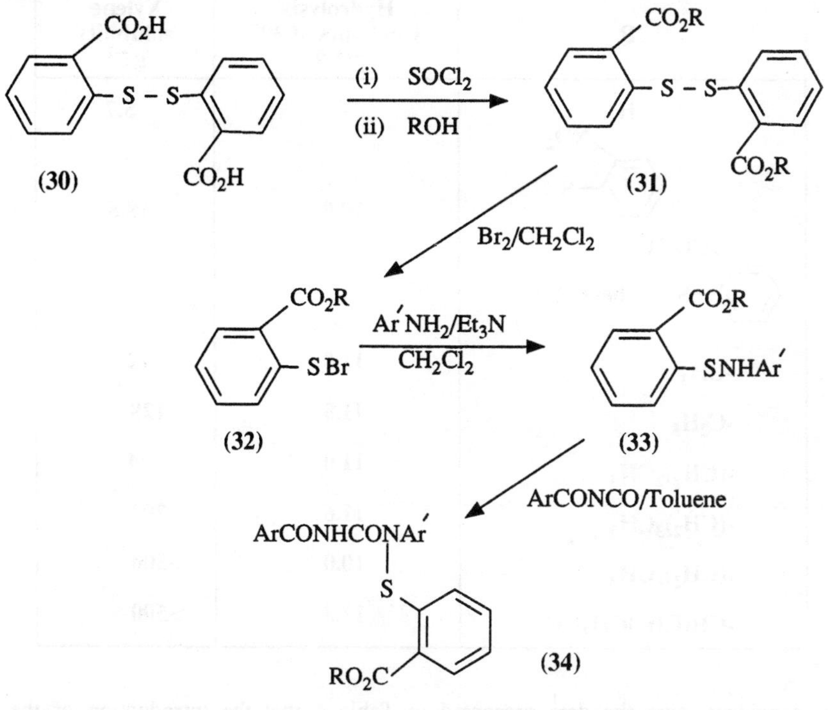

Scheme 7: Synthesis of N-(2-alkoxycarbonyl)benzenesulphenyl acylureas

The 2-(alkoxycarbonyl)benzenesulphenyl derivatives proved to be readily accessible (Scheme 7). The 2-(alkoxycarbonyl)benzenesulphenyl bromides (32) were conveniently obtained from 2,2'-dithiobisbenzoic acid (30) as outlined[11] and the preparation of the sulphenamides (33) and their condensation with aroyl isocyanates was carried out essentially as described earlier (1.3). A wide range of derivatives (34) were prepared in 40-60% overall yield by this procedure.

Table 4: The effect of the N-(2-alkoxycarbonyl)benzenesulphenyl group on physical properties

R	Hydrolysis T½ in hours at 30° pH 9	Xylene solubility g l^{-1}
H	-	5.7
S–(2-NO$_2$-phenyl)	20.0	18.8
where R' =		
-CH$_3$	11.0	42
-C$_2$H$_5$	11.8	129
-(CH$_2$)$_2$CH$_3$	11.0	44
-(CH$_2$)$_3$CH$_3$	11.6	202
-(CH$_2$)$_4$CH$_3$	10.0	>500
-CH(CH$_3$)CH$_2$CH$_3$	12.3	>500

It is evident from the data presented in Table 4 that the introduction of the N-(2-alkoxycarbonyl)benzenesulphenyl substituent has a dramatic effect on solubility and that the resulting derivatives have good hydrolytic stability. As with the sulphenyl derivatives described earlier, the spectrum of insecticidal activity in laboratory screens reflected that of the unsubstituted compounds. Derivatives of the acylurea (29) showed the best all-round activity, in some cases exceeding that of the unsubstituted compound (Table 5). All of these compounds could be readily formulated as xylene-based emulsifiable concentrates and selected examples were evaluated in the field.

Table 5: Insecticidal activity of the N-(2-alkoxycarbonyl)benzenesulphenyl derivatives of (29)

R	Toxicity Index			
	A.a.	S.l./Foliage*	H.z.	S.e.
H $\begin{smallmatrix}CO_2R'\\S\end{smallmatrix}$ where R' =	780	8790	1866	228
CH$_3$	400	4680	1300	644
C$_2$H$_5$	240	7630	975	483
n-C$_3$H$_7$	160	8530	800	329
n-C$_4$H$_9$	300	7250	975	725
-CH(CH$_3$)CH$_2$CH$_3$	250	4460	1304	3154

* Broad bean
H.z.: *Heliothis zea* (corn earworm)
S.e.: *Spodoptera exigua* (beet armyworm)

In accordance with our expectations, the liquid formulations of these sulphenyl derivatives consistently out-performed wettable powder formulations of the unsubstituted acylurea, often giving as much as 2-3 times the initial level of pest control at equivalent dose rates. Despite these promising results, further studies revealed that the true potential of the 2-(alkoxycarbonyl)benzenesulphenyl derivatives was being undermined by photolysis. Analysis of foliar deposits collected from the field showed that derivatives of this type rearrange in sunlight to products (35) which are virtually inactive[12]. (The half-life of this reaction is about 1-3 days depending on conditions).

(35)

Attempts to improve the photostability of the spray deposits by means of formulation additives were unsuccessful, so our attention turned to alternative N-sulphenyl substituents with chromophores outside the emission spectrum of sunlight. (The 2-(alkoxycarbonyl)benzenesulphenyl derivatives of (29) have an absorption band at about 310 nm in methanol).

By this time, a parallel research programme at Sittingbourne had produced a series of novel fluorinated acylureas with outstanding insecticidal and acaricidal properties[13]. This work led ultimately to the development of flufenoxuron (5)[14] and it was with this compound that we continued our study of the effects of sulphenylation.

The types of derivative examined are listed in Table 6. They were prepared essentially by the method described in Section 1.3. Further preparative details are given in the references indicated. Several of the substituents such as those in (36)-(39) have been used to modify other classes of insecticide[15], but those in (40), (41), (42) and (43) appear to be the first examples of their type to have been

Table 6: N̲-sulphenyl derivatives of flufenoxuron

Type	R	Lit. Reference
(36)	$-N\begin{smallmatrix}R_1\\CO_2R_2\end{smallmatrix}$	13, 16
(37)	$-N\begin{smallmatrix}R_1\\COR_2\end{smallmatrix}$	13
(38)	$-N\overset{(CH_2)_n}{\underset{O}{\diagdown}}$	17
(39)	$-N\begin{smallmatrix}R_1\\R_2\end{smallmatrix}$	18
(40)	$-N\overset{(CH_2)_n}{\diagdown}$ R_1O_2C	13
(41)	$\begin{matrix}R_1\\|\\-C-CO_2R_2\\|\\CO_2R_2\end{matrix}$	17
(42)	$\begin{matrix}R_1\\|\\-C-COR_2\\|\\CO_2R_3\end{matrix}$	17
(43)	$\begin{matrix}R_1\\|\\-C-COR_3\\|\\R_2\end{matrix}$	17

(R_1, R_2, and R_3 are alkyl groups; n = 1 or 2)

utilised in this way. The N-(alkylcarbamylo)sulphenyl derivatives (36) were found to have the best combination of properties for continued study. They are highly active and have good solubility properties. The substituent is cheap and easy to prepare with a wide scope for structural variation. Their hydrolytic stability is less than that of the corresponding 2-(alkoxycarbonyl)benzenesulphenyl derivatives under comparable homogeneous conditions (aqueous dioxan), with half-lives at pH 9 of a few minutes. When formulated as xylene-based emulsifiable concentrates, however, their hydrolytic stability in the aqueous emulsions produced upon dilution is quite satisfactory, with half-lives of several weeks at pH 7. They also have excellent photostability.

By varying the substituents R_1 and R_2 we were able to optimise biological activity and hydrolytic stability whilst retaining good solubility in organic solvents. The laboratory activity of two examples which emerged from this exercise are compared with flufenoxuron in Table 7.

Table 7: Relative activity of N-(alkylcarbamylo)sulphenyl derivatives of flufenoxuron against insects and mites

Compound	R	Toxicity Index		$LC_{50}(\%)$
		S.I.* A.a.		T.u.
Flufenoxuron	H	16060	1300	0.00013
(44)	$S-N \overset{tBu}{\underset{CO_2Et}{}}$	7280	620	0.00038
(45)	$S-N \overset{Me}{\underset{CO_2{}^nPr}{}}$	11070	790	0.00025

* On Chinese cabbage.

The acaricidal data presented (T.u.) refers to a life cycle assay in which freshly-hatched larvae of the red spider mite (*Tetranychus urticae*), established on foliage, are sprayed with solutions of the test compounds and left to develop through to adulthood. The LC_{50} values refer to the mortality assessed at this stage. Both of the derivatives (44) and (45) were sufficiently soluble in xylene (>200 g/l @ 25°C) to be formulated as emulsifiable concentrates.

The field performance of these formulations against insects and mites was far superior to that of comparable emulsifiable concentrates prepared from the corresponding 2-(alkoxycarbonyl)benzenesulphenyl derivatives and they were much more active than flufenoxuron formulated as a wettable powder. By this time, however, the formulation of flufenoxuron was no longer limited to wettable powders and several highly effective liquid formulations such as water-dispersible concentrates (DCs) were being developed. These consist of a solution of the acylurea in a water-miscible solvent containing suitable additives, which upon dilution in water produce extremely fine, stable dispersions (average particle size <1 µm). When emulsifiable concentrates of the derivatives (44) and (45) were compared in the field with flufenoxuron formulated as a dispersible concentrate, there was now little difference in performance. In this situation the unsubstituted acylurea has a synthetic advantage over the derivatives because it can be obtained by route (a) in Scheme 5 by the reaction of a benzamide (15) with an isocyanate (16) which is readily obtained by treatment of the corresponding aniline with phosgene. The derivatives, however, have to be prepared by route (b) which, apart from the cost of the substituent, involves the relatively more expensive benzoyl isocyanate (18). In some cases the relative costs of formulation can compensate for a difference in synthesis costs, but when all factors were taken into account, these particular derivatives were not considered to be advantageous relative to flufenoxuron.

To summarise, our studies have shown that a wide variety of N-sulphenyl acylureas can be prepared by the reaction of benzoyl isocyanates with sulphenamides. By selecting appropriate sulphenyl substituents it is possible to obtain compounds with sufficient solubility in organic solvents to be formulated as emulsifiable concentrates. The utility of such formulations then depends on the stability of the sulphenyl derivative. The 2-nitrobenzenesulphenyl acylureas have excellent hydrolytic stability but poor solubility in organic solvents. The 2-(alkoxycarbonyl)benzenesulphenyl acylureas have good hydrolytic stability and adequate solubility properties, but are photolytically unstable. The most satisfactory substituent examined was the N-(alkylcarbamylo)sulphenyl group. Derivatives of this type based on flufenoxuron gave stable emulsifiable concentrates which were highly effective in the field. Their performance, however, was no better than that of specially-developed liquid formulations of the unsubstituted acylurea.

Acknowledgement

The authors wish to thank Dr. J.F. Donnellan and his colleagues for performing the biological evaluations, Dr. A.W. McCann and his staff for their work on formulation development and Dr. K. Salisbury for his contribution to the physico-chemical studies.

References:

1.   W. Maas, R. van Hes, A.C. Grosscurt and D.H. Deul, *Chem. Pflanzenshutz*, 1981, 6, 423.

2.   A.L. Black, Y-C. Chiu, M.A.H. Fahmy and T.R. Fukuto, *J. Ag. Food Chem.*, 1973, 21, 747.

3.   A. Fry, *J. Amer. Chem. Soc.*,1953, 75, 2686.

4.   B.L. Worobey and G.R.B. Webster, *J. Chromatogr.*, 1978, 153, 423.

5.   D.N. Harpp, K. Steliou and T.H. Chan, *J. Amer. Chem. Soc.*, 1978, 100, 1222.

6.   W.E. Dennis, *J. Org. Chem.*, 1970, 35, 3253.

7.   D.N. Harpp, D.F. Mullins, J. Stelliou and I. Triassi, *J. Org. Chem.*, 1979, 44, 4196.

8.   See for example: M. Furukawa, T. Suda and S. Hayashi, *Synthesis*, 1974, 282, and references therein.

9.   This is a variation of the method described by J.H. Billman, J. Garrison, R. Anderson and B. Wolnak in, *J. Amer. Chem. Soc.*, 1941, 63, 1920, where excess of the aniline was used as base.

10. W.C. Hamilton and S.J. La Placa, *J. Amer. Chem Soc.*, 1964, 86, 2289.

11. Based on methods described in: (a) L. Katz, L.S. Karger, W. Schroeder and M.S. Cohen, *J. Org. Chem.*, 1953, 18, 1380 and (b), J.C. Grivas, *J. Org. Chem.*, 1975, 40, 2029.

12. To be described in a subsequent publication.

13. M. Anderson, *Eur. Pat. Appl.*, EP-A-161019, published Nov. 13, 1985.

14. M. Anderson, J.P. Fisher, J. Robinson and P.H. Debray, *Proc. Brit. Crop Prot. Conf. - Pests and Dis.*, 1986, 89.

15. J. Drabek and R. Neumann, "Propesticides" in *Prog. in Pestic. Biochem. and Toxicol.* (eds. D.H. Hutson and T.R. Roberts), 1985, 5, 35.

16. M. Anderson, *Eur. Pat. Appl.*, EP-A-171853, published Feb. 19, 1986.

17. M. Anderson and A.G. Brinnand, *Eur. Pat. Appl.*, EP-A-216423, published April 1, 1987.

18. M. Anderson, *Brit. Pat. Appl.*, GB 2181125 A, published April 15, 1987.

A New Class of Insecticidal Dihydropyrazoles

Richard M. Jacobson

ROHM AND HAAS COMPANY, RESEARCH LABORATORIES, 727, NORRISTOWN ROAD,
SPRING HOUSE, PA19477, USA

In the 1970s Philips-Duphar B.V. disclosed the discovery of a class of novel insecticides exemplified by PH 60-41[1],[2] and the more active PH 60-42[3]. Perhaps these compounds were related in the mind of the synthesis chemist to the chitin synthesis inhibitors analogous to PH 60-40 (diflubenzuron, Dimilin). Biology, however, showed that, while these new compounds were potent insecticides, they were neurotoxins with an entirely different mode of action than any commercial insecticides.

PH 60-40
Diflubenzuron

PH 60-41

PH 60-42

PH 60-42 has exceptionally good insecticidal activity against a broad spectrum of insects. For example, in a simple primary screen a 2.5 ppm solution kills 50% of the Mexican bean beetles (*Epilachna varivestis*) under conditions where a 100 ppm solution corresponds to a 100 g/ha application rate. Realistic field rates are usually ten times higher. Likewise 50% of Southern armyworms (*Spodoptera eridania*) succumb at 0.5 ppm. This activity impressed us. However, there was in the literature some evidence for problems with PH 60-42. Sheele[4] had reported problems with photoaromatization. Führ, Mittelstaedt, and Wieneke[5] had also reported some problems with soil degradation, giving data that equates to a soil half-life of approximately 17 months.

Intrigued by the potential of this class of insecticides, we embarked on a synthesis and testing program to improve the efficacy and improve the environmental properties. We reasoned that the photoaromatization might be blocked by disubstitution at the 4-position of the dihydropyrazole ring.

Treatment of PH 60-42 (phenyl/hydrogen at the 4 position) with two equivalents of lithium diisopropyl-amide in tetrahydrofuran created a dianion which could be alkylated with methyl iodide to yield the monomethylated derivative shown in Scheme 1. While we suffered a small loss in caterpillar activity, we were encouraged.

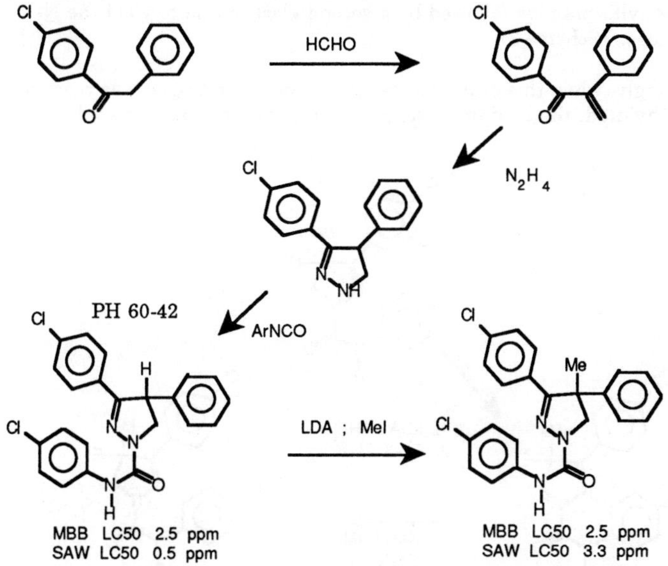
Cl
HCHO
N2H4
PH 60-42
ArNCO
H
Me
LDA ; MeI
MBB LC50 2.5 ppm
SAW LC50 0.5 ppm
MBB LC50 2.5 ppm
SAW LC50 3.3 ppm

Scheme 1

To enable a wider variety of substituents at the 4 position of the dihydropyrazole,we also made the corresponding methyl/hydrogen analog shown in Scheme 2.

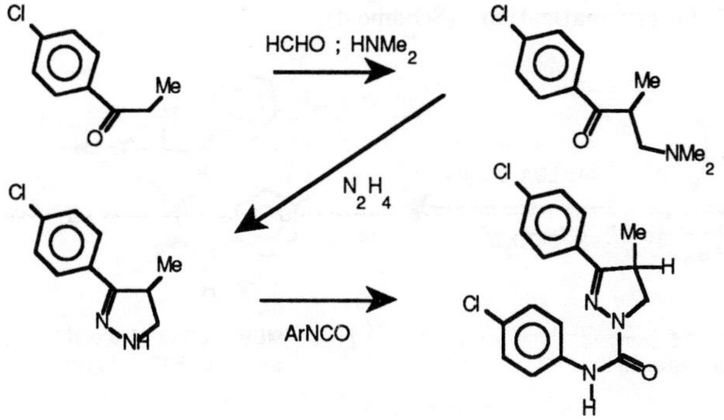
HCHO ; HNMe2
Me
NMe2
N2H4
ArNCO
Cl

Scheme 2

Now, from either the phenyl/hydrogen or the methyl/hydrogen analogs, we were able to explore reactions leading to disubstitution at the 4 position. Reaction with two equivalents of lithium diisopropylamide in tetrahydrofuran followed by addition of many different electrophiles gave the corresponding 4,4-disubstituted dihydropyrazoles. These could be further reacted using, for example, sodium hydride in dimethylformamide followed by a second electrophile to yield the $\underline{N}$-substituted analogs. (Scheme 3)

Interestingly, when this order was reversed and the nitrogen was reacted first, say $\underline{N}$-methylated, the lithium diisopropylamide deprotonation failed.

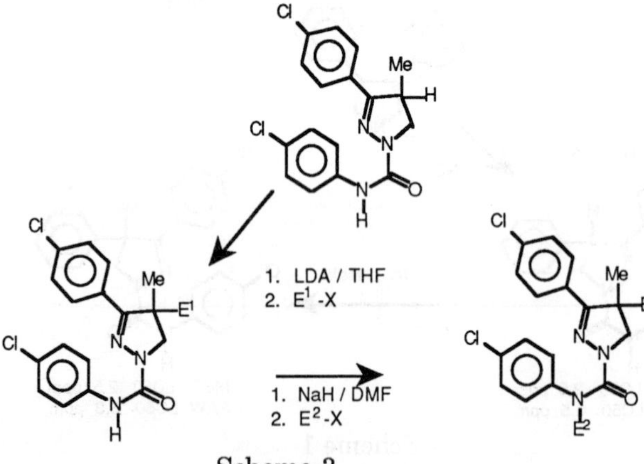

1. LDA / THF
2. E^1 -X

1. NaH / DMF
2. E^2 -X

Scheme 3

After many analogs, we discovered an especially interesting subclass of the 4,4-disubstituted dihydropyrazoles. The methyl/hydrogen compound, upon reaction with methyl chloroformate, gave a methyl/carbomethoxy compound with insecticidal activity similar to PH 60-42 but with a much shorter soil half-life, and no evidence of photoaromatization. (Scheme 4)

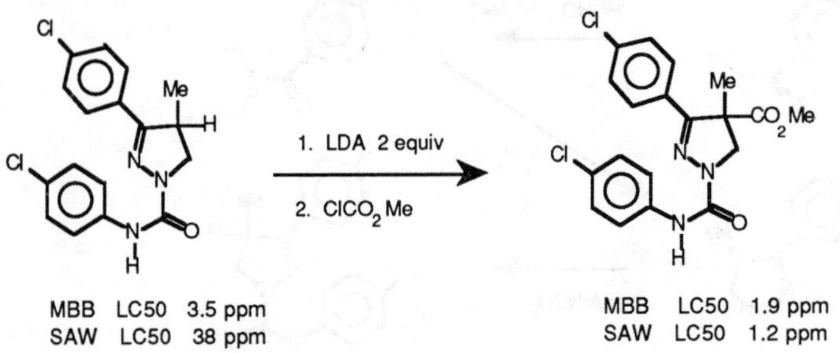

1. LDA 2 equiv

2. ClCO$_2$ Me

| MBB | LC50 | 3.5 ppm |
| SAW | LC50 | 38 ppm |

| MBB | LC50 | 1.9 ppm |
| SAW | LC50 | 1.2 ppm |

Scheme 4

Eventually a more economical route to this compound was developed (Scheme 5). Methyl 2-(4-chlorobenzoyl)propionate is readily available in high yield by a variety of routes requiring bases no stronger than sodium methoxide or aqueous caustic. It can be hydroxymethylated with formaldehyde and then mesylated with methanesulfonyl chloride to yield the mesylate in 74% yield. Treatment of the mesylate with hydrazine gives the associated dihydropyrazole, which can be carbamoylated with 4-chlorophenyl isocyanate to give the crystalline methyl/carbomethoxy compound in 95% yield over two steps. This, we believe, constitutes an economical and commercially feasible synthesis of the compound.

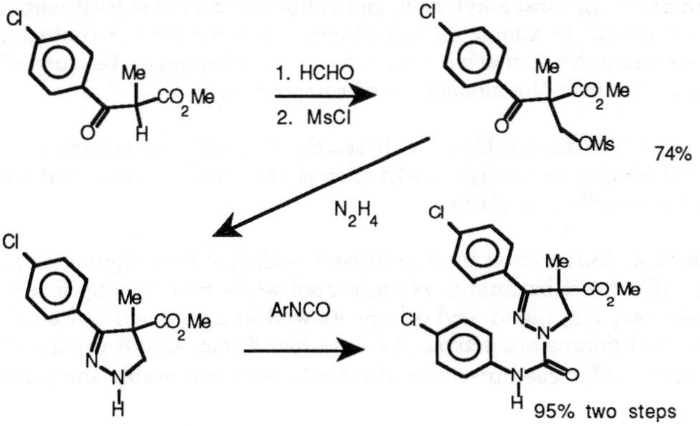

Scheme 5

Let us turn now to the structure-activity relationships of the 4,4-disubstituted dihydropyrazoles. For the purposes of this discussion, we have divided the molecule into five regions of substitution. Luckily, there seems to be little interaction between the regions. That is, an optimized substituent in one region seems to remain among the best substitutents when optimizing substituents in another region.

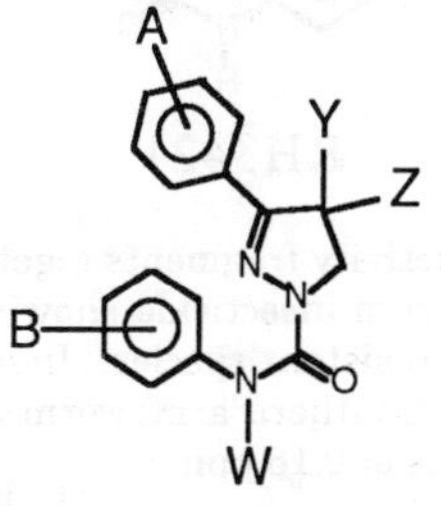

Substitution at A: The phenyl at the 3 position of the dihydropyrazole ring tolerates a wide variety of substituents while maintaining insecticidal activity. While para substitution is marginally preferred, meta substitution, ortho substitution, disubstitution, and even no substitution usually lead to active compounds. Both electron-donating and electron-withdrawing groups give compounds with good insecticidal activity. Likewise groups as large as phenyl and phenoxy are acceptable. Para-chloro substitution always seemed to be among the best.

Substitution at B: In sharp contrast to the permissiveness of the A ring, the isocyanate derived phenyl is very sensitive to substitution. Electron-donating groups, such as methoxy, greatly diminish activity. Electron-withdrawing groups, such as chloro, are acceptable but there are moderately large variations in activity with relatively small changes in structure. The position of substitution is also critical, with para substitution generally better than meta or 3,4-disubstitution. Ortho, 3,5-disubstitution, or no substitution gives loss of activity. Para-CF3 or para-OCF3 substitution were usually among the best.

Substitution at W: Several alkyl, acyl, and sulfur containing substituents are possible in this series. In almost all cases there is a diminution of contact activity, though in some cases the stomach poison activity is maintained. Because of cost, we find little advantage over the unsubstituted compounds.

Substitution at Y: We have arbitrarily chosen the Y substituent to be the smaller of the Y/Z pair. Hydrogen or methyl substitution is acceptable in this position; larger substituents are usually less active.

Substitution at Z: Variation at the Z position is really the most significant part of our work. Many different substituents are tolerated while maintaining insecticidal activity. Hydrogen, alkyl, and aryl groups as well as functionalized alkyl and functionalized aryl groups are active. We have found that -CO_2R groups and related derivatives offer good insecticidal activity with acceptably short soil half-lives.

RH3421

Putting these structure activity fragments together yields RH3421. It is an exceptionally potent broad-spectrum insecticide showing little or no cross resistance to OP, carbamate, or pyrethroid resistant species. In our simple primary screen, for example, it has an LC50 against Southern armyworm of 1.9 ppm and an impressive LC50 against Mexican bean beetle of 0.16 ppm.

We have found that RH3421 is a highly effective insecticide against representative:

Lepidoptera	-	Caterpillars
Coleoptera	-	Beetles
Diptera	-	Flies, Leafminers
Orthoptera	-	Grasshoppers, Crickets, Cockroaches
Hymenoptera	-	Ants, Bees
Homoptera	-	Planthoppers, Leafhoppers

RH3421 has less efficacy against:

Hemiptera	-	Bugs
Isoptera	-	Termites

RH3421 has little or no efficacy against:

Homoptera	-	Aphids
Thysanoptera	-	Thrips
Acarina	-	Mites
Nematoda	-	Nematodes
Mollusca	-	Snails, Slugs

The acute mammalian toxicology profile for RH3421 is very encouraging, although the compound is toxic to fish and bees:

Acute Oral (rat & rabbit)	LD50 > 5000 mpk
Acute Dermal (rat & rabbit)	LD50 > 1000 mpk
Eye Irritation	inconsequential
Skin Irritation	practically non-irritating
Mutagenicity	negative
Teratagenicity	negative

Environmental Toxicology LD50:

Fish	0.04 to 0.11 mg/L
Bird	> 500 ppm
Daphnia	0.14 mg/L
Bee	0.060 µg/bee

Soil half-life ~ 30 days

The mode of action of RH3421 has been studied and partly elucidated. RH3421 acts as a voltage-dependent sodium channel blocker. It has an anesthetic type effect on the sodium channel and is selective for cells with a low resting potential, eg. sensory neurons.[6]

In summary, the discovery of RH3421 has presented a solution to some of the problems associated with previously available dihydropyrazole insecticides. While maintaining high activity and broad spectrum, photoaromatization is no longer observed, and the soil half-life has been reduced to an acceptable 1 month. In addition, a commercial caliber synthesis has been developed.

[1] R. Mulder, K. Wellinga, and J. J. van Daalen, *Naturwissenschaften*, 1975, **62**, 531.

[2] K. Wellinga, A. C. Grosscurt, and R. van Hes, *J. Agric. Food Chem.*, 1977, **25**, 988.

[3] A. C. Grosscurt, R. van Hes, and K. Wellinga, *J. Agric. Food Chem.*, 1979, **27**, 406.

[4] B. Scheele, *Chemosphere*, **9**, 1980, 483 .

[5] F. Führ, W. Mittelstaedt, and J. Wieneke, *Chemosphere*, 1980, **9**, 469.

[6] V. Salgado, *Pesticide Science*, 1989, in press.

Trioxabicyclooctanes: GABA Receptor Binding Site and Comparative Toxicology

John E. Casida, Loretta M. Cole, Jon E. Hawkinson, and Christopher J. Palmer.

PESTICIDE CHEMISTRY AND TOXICOLOGY LABORATORY, DEPARTMENT OF ENTOMOLOGICAL SCIENCES, UNIVERSITY OF CALIFORNIA, BERKELEY, CALIFORNIA 94720, USA

1 INTRODUCTION

Pest insect control since the 1940s has been primarily dependent on three sensitive targets in the nervous system. Acetylcholinesterase (AChE) was important throughout this period with organophosphorus compounds and later methylcarbamates as inhibitors. The voltage-dependent Na^+ channel became a prominent target with DDT and maintained its importance as DDT was phased out and replaced, in part, by synthetic pyrethroids which have a similar mechanism of action. The polychlorocycloalkanes (PCCAs) including the chlorinated cyclodienes, lindane and the polychlorobornanes (i.e., toxaphene components) were used to the extent of 3×10^9 pounds before they were largely discontinued, even before the primary target of the PCCAs was defined as the $GABA_A$ receptor. The demise of DDT and the PCCAs can be attributed largely to their unacceptable environmental impact and chronic toxicity.

Knowledge gained during the past five years with bicyclophosphorus esters (BPs), bicycloorthocarboxylates (BOCs) and particularly the bicycloorthobenzoates (OBs) allows a reevaluation of the $GABA_A$ receptor as a target for insecticide action.[1-5] These trioxabicyclooctanes (TBOs) can be substituted in many ways to alter their

BPs (P = S or O)   TBPS (* also ^{35}S)

OBs   TBOB (* also³H)

receptor potency and toxicity. [^{35}S]$\underline{t}$-$\underline{B}$utylbicyclo-$\underline{p}$hosphoro$\underline{thionate}$ ([^{35}S]TBPS) and [^{3}H]$\underline{t}$-$\underline{b}$utylbicyclo-$\underline{o}$rtho$\underline{b}$enzoate ([^{3}H]TBOB) are established radioligands for the GABA$_A$ receptor/Cl$^-$ channel. The BPs, OBs, and related TBOs are used in this report involving literature review and new data to evaluate the comparative toxicology of GABA-gated Cl$^-$ channel blockers in mammals and insects.

2 DESIGNATIONS, SYNTHESES AND BIOASSAYS

BPs and OBs are referred to by abbreviations based on the 4- and 1-substituents, $\underline{e}$.$\underline{g}$. tBuPS (or TBPS) for the phosphorothionate and tBuOB (or TBOB) for the $\underline{ortho}$-benzoate. Abbreviations used for substitutents are Me methyl, Et ethyl, nPr n-propyl, nBu n-butyl, sBu s-butyl, tBu t-butyl, cHex cyclohexyl, Ph phenyl, CN cyano, HC≡C ethynyl, and TMS trimethylsilyl. Other abbreviations are THF for tetrahydrofuran and EtOAc for ethyl acetate.

Figure 1 gives the syntheses from the appropriately-substituted oxetane esters of six of the BOC probes used in studying the molecular toxicology of the GABA$_A$ receptor.

Quantitative data are given for three types of bio-assays. The receptor target is referred to as the TBO binding site since all recent studies involve a TBO radioligand, $\underline{i}$.$\underline{e}$., TBPS, TBOB or another OB. Receptor potencies at the TBO binding site are compared on the basis of nM concentrations for 50% inhibition (IC$_{50}$) of [^{35}S]TBPS or [^{3}H]TBOB binding to mammalian brain P$_2$ membranes. IC$_{50}$ values (nM) for [^{35}S]TBPS binding are 62 and 49 for TBPS and TBOB, respectively. In general similar results are obtained with either radioligand. Topical toxicity to houseflies is given as the μg/g LD$_{50}$ alone or following pretreatment with the microsomal oxidase inhibitor piperonyl butoxide (PB). Toxicity to mice is reported as the mg/kg LD$_{50}$ following intraperitoneal (IP) administration.

Figure 1. Syntheses from the oxetane esters of some of the newer bicycloorthocarboxylate probes for the $GABA_A$ receptor (see text for references). Reaction conditions: (a) $BF_3.Et_2O,CH_2Cl_2$; (b) H_2,Pd/C,EtOAc; (c) Ag_2CO_3/celite,C_6H_6; (d) 1) $TsNHNH_2,CH_2Cl_2$; 2) Al_2O_3; (e) 1) Br_2,CH_2Cl_2; 2) Et_2NH; (f) reduction; (g) $CSCl_2$; (h) $Me_3SiC{\equiv}CH$, CuI, $(Ph_3P)_2PdCl_2$, Et_2NH; (i) $nBu_4N^{\oplus}F^{\ominus}$/THF; (j) $^3H_2$, 5% Pt/C, EtOAc.

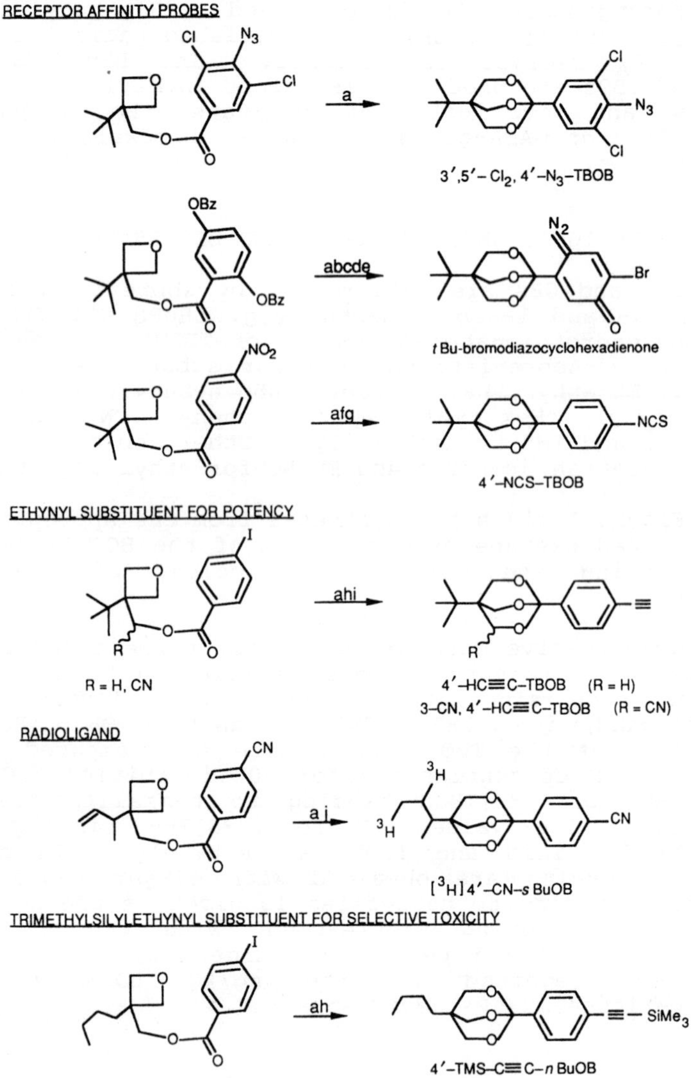
RECEPTOR AFFINITY PROBES
a
3′,5′-Cl2, 4′-N3-TBOB
abcde
t Bu-bromodiazocyclohexadienone
afg
4′-NCS-TBOB
ETHYNYL SUBSTITUENT FOR POTENCY
R = H, CN
ahi
4′-HC≡C-TBOB (R = H)
3-CN, 4′-HC≡C-TBOB (R = CN)
RADIOLIGAND
aj
[3H] 4′-CN-s BuOB
TRIMETHYLSILYLETHYNYL SUBSTITUENT FOR SELECTIVE TOXICITY
ah
4′-TMS-C≡C-n BuOB

3 PHYSIOLOGY, STRUCTURE, PHARMACOLOGY AND TOXICOLOGY OF THE GABA$_A$ RECEPTOR OF MAMMALIAN CNS

Figure 2 shows diagrammatically the relationships between GABA, the Cl⁻ channel, the TBO/insecticide site and the action of various drugs and toxicants. More detailed consideration is given below and in books on the GABA$_A$ receptor complex.[6-8]

Physiology of the GABA$_A$ Receptor (Figure 2). GABA is the major inhibitory neurotransmitter in the vertebrate central nervous system (CNS). When an inhibitory neuron is stimulated, an action potential travels down its axon and induces the release of GABA from presynaptic stores. The neurotransmitter diffuses across the synaptic cleft and binds to its recognition site on the postsynaptic GABA$_A$ receptor complex, resulting in an activation of Cl⁻ conductance. Since the electrochemical gradient for Cl⁻ is generally directed inward, increased Cl⁻ permeability induces a hyperpolarization in the postsynaptic cell, reducing the likelihood that it will fire.[6-8]

Structure of the GABA$_A$ Receptor (Figure 2). Photoaffinity labeling of the cerebral cortex GABA$_A$ receptor complex with [³H]flunitrazepam and [³H]muscimol indicates that the benzodiazepine (BZ) site is located on a 53 kD protein[10] (the α subunit) and the GABA recognition site is on a 57 kD protein[11] (the β subunit), respectively. The proposed α$_2$β$_2$ subunit structure of the purified receptor[12] is consistent with an observed molecular weight of 240 kD[13] and the presence of two gel electrophoretic bands in approximately equal amounts.[10] The amino acid sequences of the subunits (deduced from the cloned DNA sequences) and their hydropathy profiles indicate considerable structural similarity with each other and with other ion channel-containing receptors. Each subunit possesses four relatively hydrophobic regions which are thought to span the membrane and constitute the Cl⁻ channel. Co-expression of the in vitro-generated subunit RNAs in Xenopus oocytes results in functional GABA-dependent Cl⁻ channels.[12] Individually expressed subunits form GABA-stimulated Cl⁻ channels which are blocked by picrotoxinin (PTX) (discussed later), suggesting that a TBO site can be formed by either subunit.[14] A third subunit (termed γ) has been cloned which confers BZ modulation when co-expressed with both α and β subunits.[15] Since the α and γ subunits may comigrate on electrophoresis,[15] it is possible that the 53 kD subunit photoaffinity labeled by

Figure 2.   Physiology, structure, pharmacology and toxicology of
the GABA$_A$ receptor complex with emphasis on the TBO/insecticide
site of the GABA-gated Cl$^-$ channel in the mammalian CNS. Physiology
refers to the inhibitory GABAergic terminal and postsynaptic GABA$_A$
receptor complex which increases the Cl$^-$ permeability of the postsynaptic membrane. Structure involves α, β, and γ subunits each of
defined amino acid sequence and DNA coding. The two GABA and one
benzodiazepine sites per channel are preferentially located on the
β- and the α- or γ-subunits, respectively. Pharmacology involves the
interactions of drugs and toxicants with the TBO site. The effects
of various compounds on [^{35}S]TBPS binding are indicated as (-) for
inhibition and (+) for stimulation. An asterisk indicates that
GABA is required. GABA agonists (GABA and muscimol) (-) open the
channel and are blocked by competitive antagonists (bicuculline and
R 5135) (+). Benzodiazepine agonists (flunitrazepam) (-) are depres-
sants which increase the frequency of channel opening, whereas
inverse agonists (+) are convulsants; antagonists have no effect on
TBPS binding. Barbiturates (pentobarbital) (- or biphasic) increase
the mean open time of the channel. Toxicology involves convulsants
(see text) acting as channel blockers at a site possibly in the
lumen. Anticonvulsants include α-substituted-γ-butyrolactones.
Diagrams modified from references 8 and 9.

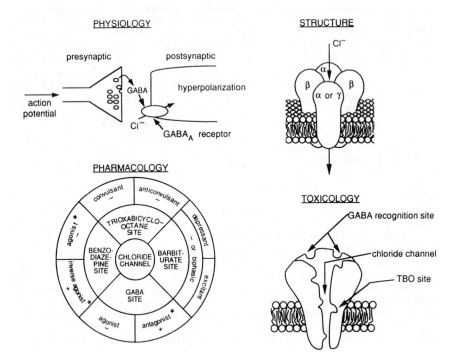

[^{3}H]flunitrazepam is in fact the γ subunit and the actual subunit structure of the receptor may be αβ$_2$γ.

Structural association of the TBO site (using [^{35}S]TBPS) with the GABA recognition site (using [^{3}H]muscimol) and the BZ receptor site (using [^{3}H]flunitrazepam) is shown by copurification of the three activities on gel filtration, density gradient centrifugation and affinity chromatography[13,16], co-immunoprecipitation[17], and coordinated developmental expression.[18]

<u>TBOs as Radioligand Probes</u>. [^{35}S]TBPS[19] is the most frequently-used radioligand for the TBO binding site of the GABA$_A$ receptor. Saturation analysis generally indicates a single population of binding sites in rat brain membranes with K_d and B_{max} values of 58 nM and 2.7 pmol/mg protein, respectively (average of 12 literature reports). However, there is also some evidence for multiple binding sites based on saturation analysis[20] and multiphasic dissociation kinetics.[19] The TBO site appears to be located at or near the anion binding site in the Cl$^-$ channel lumen based upon the correlation between the activity of various anions in stimulating [^{35}S]TBPS binding and their permeabilities through the channel in electrophysiological studies[21] and that certain organic anions block [^{35}S]TBPS binding.[22] The integrity of the TBO site requires an appropriate lipid environment as shown by phospholipase A$_2$-induced loss of [^{35}S]TBPS binding[23] and the strict dependence of binding on the presence of CHAPS in solubilized receptor preparations.[16]

[^{35}S]TBPS binding sites are localized in the cerebral cortex, cerebellum and hippocampus with lower densities in the hypothalamus, striatum and pons-medulla.[19] [^{35}S]TBPS binds to human, cow, rat, chicken, and fish brain membranes similarly[24] but significant differences have been observed in insect preparations (discussed later), indicating evolutionary divergence of the GABA$_A$ receptor between vertebrates and invertebrates.

Two other TBO radioligands are possible alternatives to [^{35}S]TBPS. The BP radioligand [^{3}H]nPrPO (structure given later) has a low affinity (K_d 30 μM) for rat brain membranes and no correlation is found between inhibition of binding by 4-substituted BPs and mammalian toxicity.[25] The binding of the OB radioligand, [^{3}H]TBOB, is similar to that of [^{35}S]TBPS with respect to pH dependence, stimulation by anions, regional distribution in brain, and *in vitro* pharmacology, but differences do exist. Unlike [35S]TBPS, [3H]TBOB binds at 0, 25 and 37°C with similar K_d values and inhibition by GABA is not affected by the NaCl concentration. Saturation analysis indicates that TBPS competitively inhibits [3H]TBOB binding.[26]

Pharmacology of the TBO Site (Figure 2). [35S]TBPS binding studies demonstrate functional association of numerous pharmacological agents with the $GABA_A$ receptor complex. GABA recognition site agonists such as muscimol and GABA noncompetitively inhibit [35S]TBPS binding[19] whereas GABA site antagonists, e.g. R 5135 and bicuculline, reverse the inhibitory effect of GABA on [35S]TBPS binding. The potencies exhibited by these GABA site ligands on [35S]TBPS binding correspond to their activities in electrophysiological and GABA site binding studies.[27]

Three classes of central BZ site ligands differ in their effects on [35S]TBPS binding. BZ agonists facilitate the action of GABA and can therefore be termed "GABA-positive". They are non-competitive inhibitors of [35S]TBPS binding in the presence of physiological levels of GABA with IC_{50}s correlated to their *in vivo* anticonvulsant potencies. The inverse agonists, which decrease the effectiveness of GABA and are termed "GABA-negative", noncompetitively enhance [35S]TBPS binding with potencies consistent with their convulsant action. The BZ antagonists have no effect.[28]

Barbiturates apparently interact with a discrete "binding site" in the receptor complex. Depressant barbiturates stimulate [35S]TBPS binding at low concentrations and inhibit at higher concentrations.[29] Other workers find only the inhibitory phase, which is either noncompetitive[30] or mixed competitive in nature.[31]

Toxicology of the TBO Site (Figure 2). Certain convulsants including PCCAs, PTX, pentylenetetrazole (PTZ), BPs, BOCs, tetramethylenedisulfotetramine (TETS) and β-substituted-γ-butyrolactones (β-GBLs) interact directly with the TBO site, resulting in blockade of GABA-dependent Cl^- transport through the channel. There is a good correlation between [35S]TBPS binding inhibition and BP toxicity,[2] OB toxicity,[2] the minimum convulsant doses of PTZ analogs[32] and the convulsant potencies of β-GBLs.[33] Saturation analysis indicates that PTX[30,34], tetrazoles[32] and β-GBLs[35] competitively inhibit [35S]TBPS binding. The α-substituted GBLs inhibit competitively[35] and their potencies as anticonvulsants correlate with their [35S]TBPS binding inhibition.[33] The GABA$_A$ recognition site antagonist R 5135 does not reverse the inhibitory action of BOCs, BPs, PTX, PTZ, or TETS[19] further indicating competitive inhibition by these convulsants.

PCCA insecticides block the GABA-gated Cl$^-$ channel. The PCCAs inhibit [35S]TBPS binding with IC$_{50}$s ranging from 36 to 1700 nM and there is a good correlation between [35S]TBPS binding inhibition and toxicity (LD$_{50}$).[36] Saturation analysis indicates that lindane, the cyclodienes endrin, 12-ketoendrin, and heptachlor epoxide, and the toxaphene component 8-Cl-B competitively inhibit [35S]TBPS binding .[36-38] Inhibition by lindane, 12-keto-endrin, and 8-Cl-B is not reversed by R 5135,[37] again indicating a competitive interaction. When a convulsant (LD$_{50}$) dose is administered IP to mice, a series of PCCAs produce an average of 62% [35S]TBPS binding inhibition in brain membranes indicating that the acute toxicity of the PCCAs is due to blockade of the GABA$_A$ receptor complex.[39] Furthermore, cyclodiene-resistance is conferred by a change in the target site since brain membranes from endrin-resistant mosquitofish have reduced affinity for [35S]TBPS.[40]

There is a good correlation between inhibition of [35S]TBPS binding by PCCAs and inhibition of ^{36}Cl$^-$ uptake in brain microsacs.[41] Inhibition of ^{36}Cl$^-$ uptake by BOCs, BPs and other cage convulsants is also correlated to inhibition of [35S]TBPS, but these compounds appear to fall on a different regression line from the PCCAs.[41] When a correlation is made between the inhibition of [35S]TBPS binding or ^{36}Cl$^-$ uptake with *in vivo* toxicity (LD$_{50}$), these convulsants fall into two groups. Type **A** compounds (PCCAs, PTX and TBOs with large 1-substituents) possess high *in vitro* potency relative to toxicity, whereas Type **B** compounds (BPs, TETS and TBOs with small 1-substituents) have low *in vitro* potency. Type **A** compounds are generally much more potent insecticides.[42]

The α-cyano (Type II) pyrethroids also stereo-specifically inhibit [35S]TBPS binding with absolute correlation between mouse intracerebroventricular toxicity and binding inhibition.[34] These pyrethroids inhibit binding with mixed competitive/noncompetitive action[34] in a GABA-dependent manner[43] with partial reversal of inhibition by R 5135,[19,37] indicating that pyrethroids interact with a closely coupled or over-lapping site. However, the pyrethroids are relatively weak inhibitors considering their *in vivo* toxicity, suggesting that the $GABA_A$ receptor is a secondary target for their excitatory action. Finally, the anthelminthic compound avermectin B_{1a} is variously reported to stimulate [^{35}S]TBPS binding at low concentrations,[44] to have a biphasic action stimulating at low but inhibiting at high concentrations,[45] or to be purely inhibitory.[46]

4 LOCALIZATION AND MOLECULAR TOPOLOGY OF THE TRIOXA-BICYCLOOCTANE/INSECTICIDE BINDING SITE OF MAMMALIAN CNS

It has been proposed that there is one TBPS binding site per $GABA_A$ receptor complex[47] and that this site is localized within the channel.[21] The assignment of the TBO site to the α, β, or γ subunit remains unknown, as does its location within the quaternary structure of the receptor complex, i.e., its association with any particular helix or amino acid sequence. These assignments require an appropriate affinity label, preferably a photoaffinity label. Three candidate photoaffinity probes vary considerably in their potency and/or their photoreactive properties. The azidophenyl compounds, 4'-N$_3$-TBOB and 3',5'-Cl$_2$-4'-N$_3$-TBOB, have the advantage of higher potency at the TBO receptor relative to the tBu-bromodiazocyclohexadienone. At a concentration of 15-times its IC_{50}, the unlabeled 4'-N$_3$-TBOB probe gives 20% photoirreversible inhibition of [^{3}H]TBOB binding. This photoinactivation is completely protectable by an appropriate TBOB analog indicating that the TBO site has been selectively modified.[48] The unlabeled dichloroazido derivative cannot be evaluated as a photoaffinity probe because it is essentially irreversible in the dark. While less potent as an inhibitor, the tBu-bromodiazocyclohexadienone has ideal photoreactivity properties and it gives 35% irreversible inhibition of [^{3}H]TBOB binding to the TBO site at 15-times its IC_{50} and this inactivation is partially protectable by TBOB.[49] The chemical affinity probe 4'-NCS-TBOB irreversibly

4'-N$_3$-TBOB (R = H)
[^{3}H]TBOB IC_{50} 150 nM
3',5'-Cl$_2$-4'-N$_3$-TBOB (R = Cl)
[^{3}H]TBOB IC_{50} 10 nM

tBu-bromodiazocyclohexadienone
[^{3}H]TBOB IC_{50} 1100 nM

4'-NCS-TBOB
[^{3}H]TBOB IC_{50} 61 nM

blocks [35S]TBPS binding by approximately 90% at 15-times its IC_{50}, but protection of this inactivation was not demonstrated.[50]

The molecular topology of the TBO site has been described on the basis of molecular modeling and structure activity of moderately potent convulsants or insecticides, *i.e.*, PTX analogs[51] and PTX, PCCAs and tBuPO.[52] The interpretability of these studies is limited since these inhibitors vary over a wide range in potency and they may not be interchangeable probes for precisely the same site. An alternate approach is to modify the structure of TBOB for optimal fit at the TBO binding site. The potency of TBOB at the TBO receptor is enhanced by up to ~25-fold by introducing a cyano substituent(s) in the 3',4' or 3 and 4' positions or an ethynyl substituent in the 3' or 4' position (Table I).

Table I. Effects of Cyano and Ethynyl Substituents on the Potency of t-Butylbicycloorthobenzoate as a GABA-gated Chloride Channel Blocker and as a Toxicant for Houseflies and Mice

| R_3 | $R_{3'}$ | $R_{4'}$ | brain membrane IC_{50}, nM | | LD_{50}, mg/kg | |
			[35S]TBPS binding	GABA-stim. $^{36}Cl^-$ flux	housefly alone (and with PB)	mouse
H	H	H	49	100	>500(23)	1.3
H	C≡N	H	5	-	>500(1.4)	0.9
H	H	C≡N	5	40	4.8(0.23)	0.06
C≡N	H	C≡N	2	-	1.2(0.09)	0.8
H	C≡CH	H	11	-	>500(12)	6.5
H	H	C≡CH	1-5	-	0.087(0.011)	0.11
C≡N	H	C≡CH	0.25	25	0.17(0.024)	0.084

Sources of data: references 1, 2, 40 and 53-55.

An additional potency increase by an order of magnitude is achieved with the combination of the 3-CN and 4'-C≡CH substituents (the potency of the individual enantiomers is not known) (Table I). The IC_{95} of ~10 nM for this compound in the [35S]TBPS binding assay with 0.25 mg of EDTA/water-dialyzed mouse brain P_2 membrane protein in 1 ml[54] is equivalent to 40 pmol/mg. This value approaches the TBPS B_{max} of 6.6 pmol/mg protein for this membrane preparation which is the total receptor concentration or the theoretical maximum potency for an irreversible inhibitor. In poisoned mice there appears to be little if any dissociation of this inhibitor from its binding site during receptor preparation and assay[54], indicating that 3-CN-4'-HC≡C-TBOB may be a "pseudo-irreversible" inhibitor. An IC_{50} of 25 nM in the GABA-stimulated Cl^- flux assay confirms the exceptional potency of this OB. Preference should be given to the most potent inhibitors in topological modeling of the TBO/insecticide binding site, i.e., to 3-CN-4'-HC≡C-TBOB.

3-CN-4'-HC≡C-TBOB
$[^{35}S]TBPS$ IC_{50} 0.25 nM

5   INSECT GABA RECEPTOR COMPLEX

The insect CNS has a GABA-gated Cl^- channel similar to the vertebrate $GABA_A$ receptor in being modulated by BZs and barbiturates but different in the lack of activity of classical GABA site antagonists. There are also differences in modulation of the channel by drugs and toxicants as determined from studies with TBO radioligands, which this brief review will emphasize.

Pharmacology of the Insect GABA Receptor. Similar to vertebrates, GABA-stimulated Cl^- conductance in locust ganglia and cockroach motor neurons is enhanced by flunitrazepam and pentobarbital.[56,57] In contrast to vertebrates, bicuculline and other GABA antagonists have virtually no effect.[56,57] Consistent with these electrophysiological studies, bicuculline does not inhibit $[^3H]GABA$ binding to cockroach CNS membranes[58] or $[^3H]$-muscimol binding to honeybee brain.[59] GABA site agonists stimulate the binding of $[^3H]$flunitrazepam to locust ganglia[60] and housefly thoraces[61] indicating functional connection of GABA and BZ binding sites. Thus, the insect CNS possesses a GABA-gated Cl^- channel, but it cannot be classified as an A type receptor due to its bicuculline insensitivity nor as a B type receptor since it is not activated by baclofen.[57]

Toxicology of the Insect GABA Receptor. As in the vertebrate CNS, PTX is an effective blocker of GABA-stimulated Cl^- conductance[56,57] and GABA-stimulated $^{36}Cl^-$ uptake[62] in the insect CNS. Endrin and lindane noncompetitively inhibit the GABA response measured in the cockroach CNS with IC_{50}s of 0.7 and 25 µM, respectively,[63] and inhibit the GABA-mediated $^{36}Cl^-$ influx into cockroach CNS microsacs.[64] Ivermectin-induced release of [^{3}H]acetylcholine from synaptosomes prepared from cockroach ganglia is inhibited by 1-1000 nM concentrations of 3-CN,4'-HC≡C-TBOB, endosulfan, dieldrin, lindane, and PTX, but not TBPS.[5,65]

TBO Radioligands for Probing the Insect GABA Receptor. Four TBO radioligands ([^{35}S]TBPS, [^{3}H]nPrPO, [^{3}H]TBOB and [^{3}H]4'-CN-sBuOB) have been utilized to study the insect GABA receptor with results dependent on the probe and the membrane preparation. [^{35}S]TBPS binds to housefly thoracic/abdominal membranes but with 3- to 4-fold lower affinity than in the mammalian system.[66,67] ThioBPs and PCCA insecticides are more effective inhibitors than BPs, TBOB, PTX, barbiturates and BZs. The inhibitory potency of 4 thioBPs parallels their injected toxicity but somewhat surprisingly no such correlation exists for 9 BPs. Generally PCCA's of high toxicity are more potent inhibitors than those of low toxicity. [^{35}S]TBPS binding in housefly thoracic/abdominal membranes differs from that in mammalian brain in that with houseflies GABA stimulates binding and inhibition by PTX and heptachlor epoxide is not competitive by saturation analysis, whereas hexobarbital is competitive.[66,67] Somewhat similar results are obtained with [^{35}S]TBPS using housefly head membranes, with PCCAs, TBPS, BPs (the inhibitory potency of 4 BPs parallels housefly toxicity), PTX and barbiturates being effective inhibitors, but GABA and BZs appear to have no effect. Isomers of the pyrethroids deltamethrin and cypermethrin inhibit with partial stereospecificity.[68] In another study, TBPS, nPrPO, TBOB and GABA only slightly inhibit total [^{35}S]TBPS binding to housefly head membranes.[69] Locust ganglion membranes differ in that [^{35}S]TBPS binding is enhanced by dieldrin, lindane, pentobarbital, a BZ, cypermethrin and GABA, whereas PTX has no effect.[70] In contrast, GABA inhibits [^{35}S]TBPS binding in cockroach ganglion membranes.[71] [^{3}H]nPrPO, also a BP probe, binds to housefly head membranes in a manner possibly unrelated to the GABA receptor. All GABA receptor agonists and antagonists (including PCCAs) active in vertebrates and invertebrates, with the exception of the BPs, have no effect on [^{3}H]nPrPO binding in this insect preparation and there is no more than a trend in correlating inhibition of binding by BPs and injected toxicity to houseflies.[69,72]

The OBs are generally much more potent insecticides than the BPs,[3] yet there is no specific binding of [^{3}H]TBOB to insect preparations.[69] [^{3}H]4'-CN-sBuOB is the most potent insecticide used thus far as a radioligand to probe the insect receptor.[5] Binding to cockroach ganglion membranes is inhibited by 3'-CN,4'-HC≡C-TBOB, dieldrin, lindane, PTX and GABA, whereas binding is not affected by TBPS or deltamethrin.[5,73] However, no combination of radioligand and insect nerve preparation adequately correlates potency at the receptor with insecticidal activity and the corresponding unlabeled ligand is the most potent inhibitor in every system examined. OBs are known with much higher insecticidal activity than 4'-CN-sBuOB and they are, in theory, potentially superior radioligands for the insect binding site.

[^{3}H]4'-CN-*s* BuOB
([^{35}S]TBPS IC$_{50}$
5 nM in mammals)

[^{3}H]*n* PrPO
([^{35}S]TBPS IC$_{50}$ 1100 nM
in mammals)

6   INSECTICIDAL ACTIVITY

The insecticidal activity of the OBs was discovered[1] when 4'-Cl-TBOB, prepared as a precursor for the radioligand [^{3}H]TBOB,[26] was tested for topical toxicity to houseflies and cockroaches. The interesting level of potency prompted synthesis of its 4-nPr, 4-tBu and 4-cHex analogs with other 4'-halogen substituents (Table II). The halogen substituents, particularly bromo and iodo, increase the potency of TBOB so that significant insecticidal activity is then evident without PB. A 4'-cyano group is one of the few substituents as effective as the best 4'-halogen in conferring housefly toxicity[1] (Tables I and II).

The insecticidal activity of TBOB is remarkably enhanced by appropriately positioned cyano and ethynyl substituents (Table I).   The potency conferred by cyano substituents increases in the order 3'-CN < 4'-CN < 3-CN,4'-CN.   Relative to the cyano group, the ethynyl substituent is less effective in the 3'-position but more effective in the 4'-position and the addition of the 3-cyano group does not increase the potency of the 4'-ethynyl compound as it does for its 4'-cyano analog. The

Table II. Effects of 4'-Halogen Substituents on the Potency of 4-Alkyl Bicycloorthobenzoates as Trioxabicyclooctane Receptor Inhibitors and as Selective Toxicants for Houseflies and Mice

X substituent of R-C(CH$_2$O)$_3$C-Ph-4-X

species and R substituent	H	F	Cl	Br	I
			[^{35}S]TBPS IC$_{50}$, nM		
Mammalian brain					
nPr	1500	-	176	-	136
tBu	49[a]	42	7[b]	10	12
cHex	41	21	13	19	55
			LD$_{50}$ alone (and with PB), mg/kg		
Housefly					
nPr	>500(90)	-	23(2.5)	-	5.4(1.7)
tBu	>500(23)[a]	>500(5.5)	10(1.5)[b]	3.5(0.83)	4(1.4)
cHex	>500(13)	>500(1.9)	10(0.53)	6.5(0.25)	3.3(0.27)
Mouse					
nPr	10	-	9.1	-	>100
tBu	1.3[a]	0.77	1.1[b]	1.2	6
cHex	5.6(2.3)	3.5(1.2)	52(5.0)	4.8	>33
			Selectivity (LD$_{50}$ mouse/LD$_{50}$ housefly)		
nPr	<0.02	-	0.39	-	>19
tBu	<0.003[a]	<0.002	0.11[b]	0.34	1.5
cHex	<0.01	0.007	5.2	0.74	>10

[a]TBOB   [b]4'-Cl-TBOB
Sources of data: references 1, 2 and 55.

high potency of 4'-HC≡C-TBOB applies to American cockroaches as well as to houseflies.[53]

4'-HC≡C-TBOB
([^{35}S]TBPS IC_{50} 1·5 nM
in mammals)

The outstanding insecticidal activity of 4'-HC≡C-TBOB is evident by comparing its potency on houseflies to those of several established insecticides with various modes of action (Table III). 4'-HC≡C-TBOB is much more potent than dieldrin, also a GABA-gated Cl$^-$ channel blocker. It is less toxic than [1R, cis, αS]cypermethrin at 25° C but more toxic at 35° C, due to a difference in the temperature coefficients of insecticides acting at the GABA-gated Cl$^-$ channel and the voltage-dependent Na$^+$ channel. The positive temperature coefficient of the GABA-gated Cl$^-$ channel blockers is often an advantage under ambient environmental conditions. DDT, also acting at the Na$^+$ channel, is much less active as are the AChE inhibitors parathion and propoxur. Thus, 4'-HC≡C-TBOB approximates or exceeds the potency of the most effective insecticides acting on other targets (Table III). This clearly establishes that the sensitivity and toxicological relevance of the insect GABA receptor, at least of houseflies and American cockroaches, are not limiting factors in designing potent insecticides acting at this target.

7 SELECTIVE TOXICITY

The first OBs examined, i.e., TBOB and 4'-Cl-TBOB, are selectively toxic to mice compared with houseflies. In comparing the 4-nPr, 4-tBu and 4-cHex compounds, the tBu analogs are most toxic to mice and tend to be most potent at the TBO receptor (Table II). Higher toxicity for houseflies than mice is evident with 4'-Cl-cHexOB, 4'-I-nPrOB, 4'-I-TBOB and 4'-I-cHexOB (Tables II and IV).

The 4'-cyano and 4'-substituted-ethynyl compounds are particularly interesting relative to selective toxicity. The cyano and ethynyl analogs vary from equitoxic to mice and houseflies (4'-HC≡C-TBOB and 3-CN, 4'-CN-TBOB) to much more toxic to mice than to houseflies (e.g., 4'-CN-TBOB) (Table I). The 4'-(substituted-ethynyl) derivatives, R_4-C(CH$_2$O)$_3$C-Ph-C≡C-R',

Table III. Toxicity to Houseflies of 4'-Ethynyl-t-butylbicyclo-orthobenzoate Acting at the GABA-Gated Chloride Channel Relative to Those of Other Classes of Insecticides Acting at the Voltage-Dependent Sodium Channel and Acetylcholinesterase

insecticide	temp. °C	LD_{50}, $\mu g/g$ alone	LD_{50}, $\mu g/g$ with PB	mode of action
4'-HC≡C-TBOB	35	0.064	0.009	Cl^-
	25	0.087	0.011	Cl^-
dieldrin	25	0.65	0.83	Cl^-
[1R,cis,αS]-cypermethrin	35	0.14	0.015	Na^+
	25	0.029	0.0012	Na^+
DDT	25	14	12	Na^+
parathion	25	1.3	0.43	AChE
propoxur	25	23	1.4	AChE

Source of data: reference 53.

Table IV. Effects of 4- and 4'-Substituents on the Selective Toxicity of Bicycloorthobenzoates to Mice and Houseflies

R_4-C(CH$_2$O)$_3$C-Ph-R_4'

R_4	R_4'	LD_{50}, mg/kg mouse	LD_{50}, mg/kg housefly	LD_{50} ratio mouse/housefly
tBu	H	1.3	>500	<0.003
tBu	CN	0.06	4.8	0.013
nPr	C≡CCH(OEt)$_2$	0.34	23	0.015
tBu	Cl	1.1	10	0.11
cHex	Cl	52	10	5.2
nPr	I	>100	5.4	>19
nBu	C≡C-TMS	>400	0.43	>930

Sources of data: references 1 and 74.

show remarkable substituent effects on selective toxicity (Table IV). The diethylacetal [R_4=nPr, R'=CH(OEt)$_2$] is 68-fold more toxic to mice than to houseflies. In contrast the TMS compound (R_4=nBu, R'=SiMe$_3$) is >930-fold more toxic to houseflies than to mice.[74] It is therefore clear that GABA-gated Cl$^-$ channel blockers may be selectively toxic to either mammals or insects and that high levels of selective toxicity can be achieved.

8 METABOLIC DETOXIFICATION

The OBs generally undergo rapid oxidative detoxification in houseflies based on their magnitude of synergism by PB. The cyano- and ethynyl-substituted TBOBs are synergized by 7- to >357-fold (Table I). Oxidative detoxification, determined as PB synergism, is less important with 4'-Cl, 4'-Br and 4'-I compounds relative to the 4'-F and unsubstituted TBOB analogs (Table V). Compounds with the 4-cHex substituent are more readily synergized than those with the 4-nPr and 4-tBu groups (Table V) suggesting that the cHex substituent confers more rapid oxidative detoxification.

The retarding effect of 4'-Br and 4'-I substituents on TBOB metabolism, suggested by PB synergism, is also evident in the mouse liver microsome-NADPH system as analyzed by gas chromatography to determine substrate loss (Table V). The greater ease of detoxifying the 4-cHex compounds relative to the 4-tBu analogs is confirmed by studies with mouse microsomes comparing metabolism with and without the oxidase cofactor NADPH in which detoxification is evident for 4'-Cl-cHexOB under conditions in which 4'-Cl-TBOB is not detoxified (Table VI).

There are often multiple sites of metabolic attack complicating the interpretation of substituent effects on the persistence of the parent compound. For example, TBOB and 4'-CN-sBuOB undergo oxidative attack on an O-methylene substituent leading to spontaneous cage opening followed by enzymatic reduction of the resultant monoaldehyde to the monoester diol.[75,76] Although initiated by an oxidative process, this gives an overall effect comparable to hydrolytic cleavage of the OB to the monoester diol. As noted earlier with TBOB,[75] the 4-alkyl substituent of 4'-CN-sBuOB also undergoes oxidative attack, possibly involving each carbon in the sec-butyl group.[76]

Table V. Effects of 4'-Halogen Substituents on Oxidative Detoxification of 4-Alkyl-bicycloorthobenzoates

species and R substituent	X substituent of $R-C(CH_2O)_3C-Ph-X$				
	H	F	Cl	Br	I
Housefly synergism factor (LD_{50} alone/LD_{50} with PB)					
nPr	>5.5	-	9	-	3.2
tBu	>22	>91	7	4	2.9
cHex	>38	>263	19	26	12
Mouse liver microsomes, % metabolism					
tBu	17	11	14	3	2

Sources of data: Table II and reference 55.

Table VI. Effects of Microsomal Oxidases on the Potency of Three Bicycloorthobenzoates as Trioxabicyclooctane Receptor Inhibitors

R_4	R_4'	[nM]	% inhibition $[^{35}S]$TBPS binding	
			control	oxidase
tBu	Cl	7	29	34
cHex	Cl	13	50	11
nPr	C≡CCH(OEt)$_2$	1000	7	26

Sources of data: references 2 and 55.

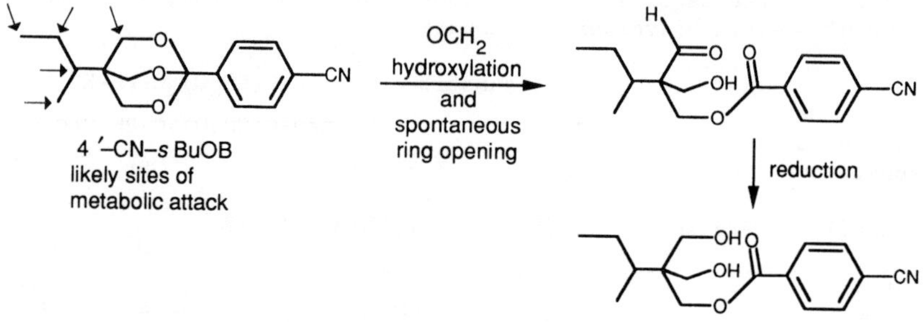

4′-CN-s BuOB
likely sites of
metabolic attack

9   METABOLIC ACTIVATION

Two examples are cited above of unusual selective toxicity for TBOB analogs with substituted-ethynyl groups (Table IV). The diethylacetal is a weak insecticide and is not active at the TBO receptor ([^{35}S]TBPS IC$_{50}$ > 10,000 nM) but is very toxic to mice.[55] On incubation with a liver microsomal oxidase system it undergoes NADPH-dependent activation to a more potent TBO receptor ligand (Table VI). It appears that the diethylacetal readily undergoes oxidative metabolic activation in mice.

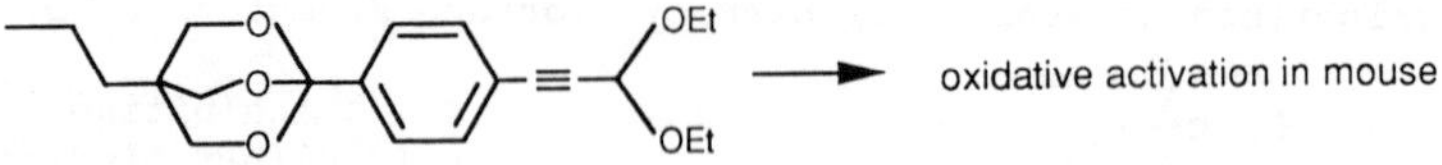

oxidative activation in mouse

In contrast, the trimethylsilylethynylphenyl derivative is very potent to houseflies with a toxicity about half that of its ethynylphenyl analog. PB strongly antagonizes the activity of the trimethylsilylethylphenyl derivative but synergises that of the ethynylphenyl compound. These findings suggest that the trimethyl-silylethynylphenyl group undergoes oxidative metabolic activation to the ethynylphenyl substituent. 4′-TMS-C≡C-nBuOB appears to have the appropriate combination of substituents for metabolic activation in houseflies but not mice and thus is a highly selective insecticide (Table VII).[74]

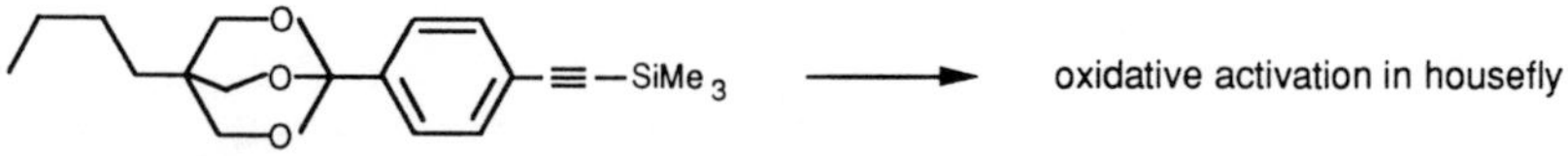

oxidative activation in housefly

4′-TMS–C≡C–*n* BuOB
[^{35}S]TBPS IC$_{50}$ 1260 nM

Table VII. Effects of 4'-Ethynyl and 4'-Trimethylsilylethynyl Substituents on the Potency of 4-n-Butylbicycloorthobenzoate as a Toxicant for Houseflies and Mice

	$nBu-C(CH_2O)_3C-Ph-C\equiv C-R$	
assay	R=H	R=SiMe$_3$
Housefly topical LD$_{50}$, μg/g		
Alone	0.24	0.43
With PB	0.054	124
Mouse IP LD$_{50}$, mg/kg	1.1	>400

Source of data: reference 74.

10 SUMMARY

The GABA-gated Cl$^-$ channel is very sensitive to channel blockers acting at the TBO/insecticide binding site. Radioligands based on potent BP and OB toxicants served as the basis for defining the properties of coupled sites in the GABA$_A$ receptor complex of the mammalian CNS. Progress towards localization of the binding site is being made with TBO affinity ligands. Emerging knowledge on the comparable insect system indicates a general similarity to the mammalian receptor although an optimal radioligand probe for the insect TBO site has yet to be developed. The most potent inhibitors are appropriately-substituted OBs which now rival the potency of established insecticides acting at other nerve targets. Examples are given of remarkable selective toxicity of GABA$_A$ receptor antagonists achieved by differential metabolic detoxification and activation.

ACKNOWLEDGMENT

We thank our University of California colleagues Hsi Liu and Judith Engel for performing the bioassays. Malcolm Black, John Larkin, Ian Smith and John Weston of Wellcome Research Laboratories (Berkhamsted, England) provided encouragement and useful discussions. This study was supported in part by National Institute of Environmental Health Sciences Grant P01 ES00049.

REFERENCES

1. C.J. Palmer and J.E. Casida, J. Agric. Food Chem., 1985, 33, 976.
2. J.E. Casida, C.J. Palmer, and L.M. Cole, Mol. Pharmacol., 1985, 28, 246.
3. C.J. Palmer and J.E. Casida, In 'Sites of Action for Neurotoxic Pesticides', R.M. Hollingworth and M.G. Green, Eds., ACS Symposium Series 356, American Chemical Society, Washington, D.C., 1987, p. 71.
4. J.E. Casida and C.J. Palmer, In 'Chloride Channels and Their Modulation by Neurotransmitters and Drugs', Raven, New York, 1988, p. 109.
5. J.E. Casida, R.A. Nicholson, and C.J. Palmer, In 'Neurotox '88: Molecular Basis of Drug & Pesticide Action', G.G. Lunt, Ed., Elsevier, Amsterdam, Netherlands, 1988, p. 125.
6. R.F. Squires, Ed., 'GABA and Benzodiazepine Receptors', CRC Press, Boca Raton, FL, Vols. I and II, 1988.
7. R.W. Olsen and J. C. Venter, Eds., 'Benzodiazepine/GABA Receptors and Chloride Channels: Structural and Functional Properties', Alan R. Liss, Inc., New York, 1986.
8. G. Biggio and E. Costa, Eds., 'Chloride Channels and Their Modulation by Neurotransmitters and Drugs', Raven, New York, 1988.
9. M. Eldefrawi and A. Eldefrawi, In 'Neurotox '88: Molecular Basis of Drug & Pesticide Action', G.G. Lunt, Ed., Elsevier, Amsterdam, Netherlands, 1988, p. 207.
10. C. Mamalaki, F.A. Stephenson, and E.A. Barnard, EMBO J., 1987, 6, 561.
11. S.O. Casalotti, F.A. Stephenson, and E.A. Barnard, J. Biol. Chem., 1986, 261, 15013.
12. P.R. Schofield, M.G. Darlison, N. Fujita, D.R. Burt, F.A. Stephenson, H. Rodriguez, L.M. Rhee, J. Ramachandran, V. Reale, T.A. Glencorse, P.H. Seeburg, and E.A. Barnard, Nature, 1987, 328, 221.
13. C. Mamalaki, E.A. Barnard, and F.A. Stephenson, J. Neurochem., 1989, 52, 124.
14. L.A.C. Blair, E.S. Levitan, J. Marshall, V.E. Dionne, and E.A. Barnard, Science, 1988, 242, 577.
15. D.B. Pritchett, H. Sontheimer, B.D. Shivers, S. Ymer, H. Kettenmann, P.R. Schofield, and P.H. Seeburg, Nature, 1989, 338, 582.
16. R.G. King, M. Nielsen, G.B. Stauber, and R.W. Olsen, Eur. J. Biochem., 1987, 169, 555.
17. J. Vitorica, D. Park, G. Chin, and A.L. de Blas, J. Neurosci, 1988, 8, 615.
18. A. Batuecas, A. Cubero, A. Barat, and G. Ramirez, Neurochem. Int., 1987, 11, 425.
19. R. F. Squires, J.E. Casida, M. Richardson, and E. Saederup, Mol. Pharmacol., 1983, 23, 326.

20. M.H.J. Tehrani, C.J. Clancey, and E.M. Barnes, Jr., J. Neurochem, 1985, 45, 1311.
21. J.C.G. Marvizon and P. Skolnick, J. Neurochem, 1988, 50, 1632.
22. G. Evoniuk and P. Skolnick, Mol. Pharmacol., 1988, 34, 837.
23. H. Havoundjian, R.M. Cohen, S.M. Paul, and P. Skolnick, J. Neurochem., 1986, 46, 804.
24. L. M. Cole, L.J. Lawrence, and J.E. Casida, Life Sci., 1984, 35, 1755.
25. Y. Ozoe and M. Eto, In 'Membrane Receptors and Enzymes as Targets of Insecticidal Action', J.M. Clark and F. Matsumura, Eds., Plenum, New York, 1986, p. 75.
26. L.J. Lawrence, C.J. Palmer, K.W. Gee, X. Wang, H.I. Yamamura, and J. E. Casida, J. Neurochem., 1985, 45, 798.
27. R.F. Squires and E. Saederup, Brain Res., 1987, 414, 357.
28. K.W. Gee, L.J. Lawrence, and H.I. Yamamura, Mol. Pharmacol., 1986, 30, 218.
29. S. Liljequist and B. Tabakoff, Life Sci., 1986, 39, 851.
30. Y. Ito, D.K. Lim, T. Nabeshima, and I.K. Ho, J. Neurochem, 1989, 52, 1064.
31. R.R. Trifiletti, A.M. Snowman, and S.H. Snyder, Eur. J. Pharmacol., 1984, 106, 441.
32. R. Ramanjaneyulu and M. K. Ticku, Eur. J. Pharmacol., 1984, 98, 337.
33. B.A. Weissman, T.R. Burke, Jr., K.C. Rice, and P. Skolnick, Eur. J. Pharmacol, 1984, 105, 195.
34. L.J. Lawrence and J.E. Casida, Science, 1983, 221, 1399.
35. J.A. Levine, J.A. Ferrendelli, and D.F. Covey, Biochem. Pharmacol., 1985, 34, 4187.
36. L.J. Lawrence and J. E. Casida, Life Sci., 1984, 35, 171.
37. J.E. Casida and L.J. Lawrence, Environ. Health Perspec., 1985, 61, 123.
38. I.M. Abalis, M.E. Eldefrawi, and A. T. Eldefrawi, Pestic. Biochem. Physiol., 1985, 24, 95.
39. L.M. Cole and J.E. Casida, Life Sci., 1986, 39, 1855.
40. J.C. Bonner and J.D. Yarbrough, Pestic. Biochem. Physiol., 1987, 29, 260.
41. T. Obata, H.I. Yamamura, E. Malatynska, M. Ikeda, H. Laird, C.J. Palmer, and J.E. Casida, J. Pharmacol. Exp. Ther., 1988, 244, 802.
42. C.J. Palmer and J. E. Casida, Toxicol. Lett., 1988, 42, 117.
43. J. Seifert and J.E. Casida, Eur. J. Pharmacol., 1985, 115, 191.
44. G. Drexler and W. Sieghart, Neurosci. Lett., 1984, 50, 273.
45. R. W. Olsen and A. M. Snowman, J. Neurochem., 1985, 44, 1074.
46. I. M. Abalis, A.T. Eldefrawi, and M.E. Eldefrawi, J. Biochem. Toxicol., 1986, 1, 69.
47. F.A. Stephenson, J. Receptor Res., 1987, 7, 43.
48. J.E. Hawkinson, M.P. Goeldner, C.J. Palmer, and J. E. Casida, unpublished results.
49. M.P. Goeldner, J.E. Hawkinson, and J.E. Casida, Tetrahedron Lett., 1989, 30, 823.

50. A.H. Lewin, B.R. de Costa, K.C. Rice, and P. Skolnick, Mol. Pharmacol., 1989, 35, 189.
51. W.E. Klunk, P.A. Kalman, J.A. Ferrendelli, and D.F. Covey, Mol. Pharmacol., 1983, 23, 511.
52. Y. Ozoe and F. Matsumura, J. Agric. Food Chem., 1986, 34, 126.
53. C.J. Palmer and J. E. Casida, J. Agric. Food Chem., 1989, 37, 213.
54. C.J. Palmer, L.M. Cole, and J. E. Casida, J. Med. Chem., 1988, 31, 1064.
55. C.J. Palmer, L.M. Cole, and J. E. Casida, unpublished results.
56. G. Lees, D.J. Beadle, R. Neumann, and J.A. Benson, Brain Res., 1987, 401, 267.
57. D.B. Sattelle, R.D. Pinnock, K.A. Wafford, and J.A. David, Proc. R. Soc. Lond., 1988, B 232, 443.
58. S.C.R. Lummis and D.B. Sattelle, Neurosci. Lett., 1985, 60, 13.
59. I.M. Abalis and A.T. Eldefrawi, Pestic. Biochem. Physiol., 1986, 25, 279.
60. T. Robinson, D. MacAllan, G. Lunt, and M. Battersby, J. Neurochem., 1986, 47, 1955.
61. A.T. Eldefrawi, I. M. Abalis, and M.E. Eldefrawi, In 'Membrane Receptors and Enzymes as Targets of Insecticidal Action', J.M. Clark and F. Matsumura, Eds., Plenum, New York, 1986, p. 107.
62. K.A. Wafford, D.B. Sattelle, I. Abalis, A.T. Eldefrawi, and M.E. Eldefrawi, J. Neurochem., 1987, 48, 177.
63. K.A. Wafford, D.B. Sattelle, D.B. Gant, A.T. Eldefrawi, and M.E. Eldefrawi, Pestic. Biochem. Physiol., 1989, 33, 213.
64. K.A. Wafford, S.C.R. Lummis, and D.B. Sattelle, Pestic. Sci., 1988, 24, 338.
65. R.A. Nicholson, P.S. Robinson, C.J. Palmer, and J.E. Casida, Pestic. Sci., 1988, 24, 185.
66. E. Cohen and J.E. Casida, Life Sci., 1985, 36, 1837.
67. E. Cohen and J.E. Casida, Pestic. Biochem. Physiol., 1986, 25, 63.
68. R.W. Olsen, O. Szamraj, and T. Miller, J. Neurochem., 1989, 52, 1311.
69. Y. Ozoe, K. Mochida, T. Nakamura, and M. Eto, Comp. Biochem. Physiol., 1988, 91C, 365.
70. M.C.S. Brown, G.C. Lunt, and A. Stapleton, Comp. Biochem. Physiol., 1989, 92C, 9.
71. S.C.R. Lummis and D.B. Sattelle, Neurochem. Int., 1986, 9, 287.
72. Y. Ozoe, M. Eto, K. Mochida, and T. Nakamura, Pestic. Biochem. Physiol., 1986, 26, 263.
73. R.A. Nicholson, C.J. Palmer, and J.E. Casida, Pestic. Sci., 1988, 24, 183.
74. C.J. Palmer, I.H. Smith, M.D.V. Moss, and J.E. Casida, in press.
75. J.G. Scott, C.J. Palmer, and J. E. Casida, Xenobiotica, 1987, 17, 1085.
76. Y.-L. Deng, C.J. Palmer, R.F. Toia, and J.E. Casida, in press.

Molecular Control of Behaviour and Gene Expression in Insects: How Pheromones and Hormones Work

G. D. Prestwich

DEPARTMENT OF CHEMISTRY, STATE UNIVERSITY OF NEW YORK, STONY BROOK, NEW YORK 11794-3400, USA

1 INTRODUCTION

The biological instructions contained in small lipid molecules, e.g., pheromones and hormones, require protein receptors to translate chemical information into specific biochemical events. Pheromones interact with peripheral sensory neurons, converting chemical binding to electrical signals which produce a change in behavior. Juvenile hormones and their analogs activate DNA-binding nuclear proteins in target tissues to suppress the changes in gene expression which occur during metamorphosis.

In order to study binding, catabolic and receptor proteins involved in the molecular action of pheromones and hormones, we have prepared high specific activity radioligands. Furthermore, chemically reactive analogs and photoaffinity labels have been synthesized for covalent modification of proteins. Recent results with the lepidopterans *Lymantria*, *Heliothis*, *Antheraea*, *Orgyia*, and others, will illustrate successful strategies in understanding pheromone biochemistry. Similarly, characterization of receptors for JH and JHA in *Manduca sexta* will be described. Finally, new efforts directed at photolabeling the receptors for peptide hormones are discussed.

The chemical tools and biological insights which emerge from these basic studies can assist in development of pest control strategies in two principal ways. First, the targeted reagents themselves may offer new lead structures for optimization of olfactory or hormonal disruption. Second, equipped with a more intimate knowledge of the molecular action of hormones, pheromones

and their chemical analogs, specific steps in the
transductory pathways can be interrupted.

2 PHEROMONE BIOCHEMISTRY

    Based on our current biochemical model for pheromone
action,[1] we have selected three types of sensillum-specific
proteins as targets for chemical study. First, soluble
pheromone binding proteins (PBP) appear to mediate
transport of pheromone molecules from the pore tubules
penetrating the cuticle to the membrane of the dendrite.
These proteins can be photoaffinity labeled and are
available in sufficient quantities for microsequencing by
gas-phase Edman degradation. Second, enzymes which degrade
pheromones are found in antennae and in other body regions.
These can be visualized in polyacrylamide gels by affinity
labeling and enzyme stains. Third, a receptor protein (RP)
has been identified in the dendritic membrane by
photoaffinity labeling.

    Recent results with lepidopteran pheromone perception
are summarized below. Several chemically distinct
approaches to selective modification of olfaction or
olfactory proteins are described according to the chemical
functional groups and insects involved.

Diazoacetate Photoaffinity Label: *Antheraea polyphemus*

    Long-chain unsaturated acetates are common
constituents of moth pheromones. As a probe for PBPs and
RPs in wild silk moth antennae, we employed [11,12-^{3}H]-
(E,Z)-6,11-hexadecadienyl diazoacetate, a photolabile
analog of the acetate pheromone. Using purified sensillum
lymph and sensory dendritic membranes of *A. polyphemus,* we
covalently modified a 15-kD soluble protein and a 69-kD
membrane protein.[2] Radioligand-modified proteins are
separated by electrophoresis and detected by fluorography.
These two proteins were unique to male antennae, and
labeling could be reduced by co-incubation with a 100-fold
excess of the acetate pheromone during the irradiation at
254 nm. These experiments constitute the first direct
evidence for both soluble pheromone-specific PBPs and
neural RPs in an insect chemosensory cell. Recently,
N terminal amino acid sequences have been obtained for both
photoaffinity-labeled proteins. Ongoing efforts are
directed at determination of covalently modified amino acid
residue(s).

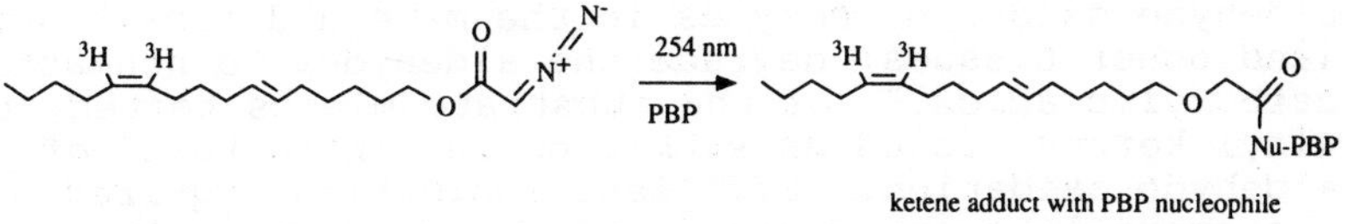
254 nm
PBP
Nu-PBP
ketene adduct with PBP nucleophile

Figure 1 Photoaffinity labeling of *A. polyphemus* PBP

Diazoketone Photoaffinity Label: *Orgyia pseudotsugata*

Ketones are relatively uncommon functionalities in pheromones. We selected the Douglas fir tussock moth *Orgyia pseudotsugata*, a lymantriid cousin of the gypsy moth, to examine PBPs for ketonic pheromones. First, we synthesized [6,7-^{3}H]-*(Z)*-6-heneicosen-11-one by catalytic semitritation of the 6-alkyne. We were surprised to observe two sets of AB quartets for the vinyl tritons in the 320 MHz tritium NMR spectrum.[3] The anisochronous vinyl signals for the *(Z)* isomer (85%) show J_{AB} = 12.4 Hz, while the pattern for the *(E)* isomer (15%) showed a 16.7 Hz coupling constant.

The ketone was converted to a physiologically and photolabile diazoketone, and irradiation with antennal proteins at 254 nm resulted in covalent modification of a protein band at ca. 18 kD. However, we could not demonstrate displacement by excess pheromone. It is noteworthy that this is a general problem with diazoketone photoaffinity labels, as discussed below for methoprene diazoketone. Furthermore, we could not detect any metabolism of the [^{3}H]-ketone *in vitro* using conditions in which >50% turnover was observed for silk moth and gypsy moth assays.

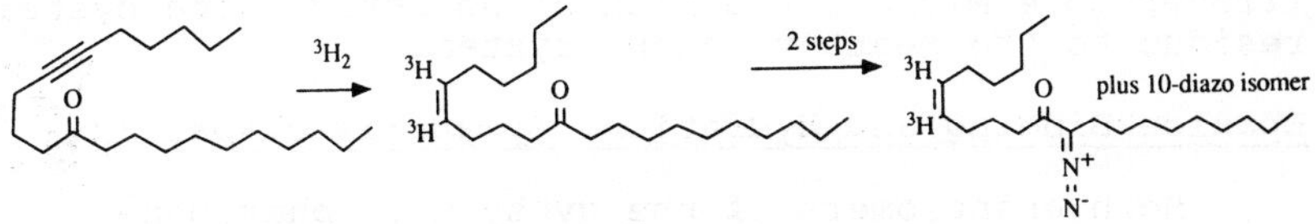
2 steps
plus 10-diazo isomer

Figure 2 Synthesis of a diazoketone photolabel for *Orgyia*

Vinyl Ketones as Affinity Labels: *Heliothis virescens*

The tobacco budworm moth, *Heliothis virescens*, uses a seven-component blend of aldehydes as the sex pheromone. A primary alcohol-specific, hexane-stable alcohol oxidase produces these aldehydes in the female glands,[4] and

aldehyde oxidizing enzymes in the male and female antennae
(and other tissues) degrade the aldehydes to non-attractive
carboxylic acids.[5] Of the substrate mimics tested, two
vinyl ketones acted as submicromolar inhibitors[6] of
aldehyde oxidation. Efficient inhibition required a
pheromone-like structure,[7] and these two materials showed
weak electrophysiological activity (M. Giblin and
E. Underwood, personal communication). However, despite
the potent enzyme inhibition, we failed to obtain evidence
that these reactive pheromone analogs had any effect
in vivo on pheromone perception.

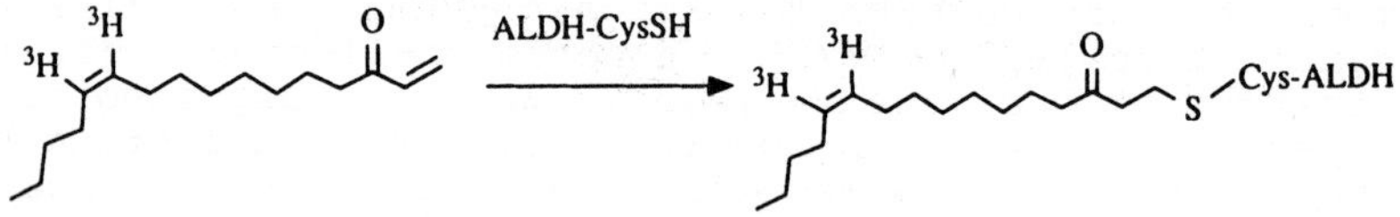

Figure 3   Affinity labeling of ALDH with [3H]-vinyl ketone

Enzyme-specific staining of native PAGE-separated
proteins from a variety of tissues showed a family of
antennal-specific aldehyde oxidase (AO) isozymes which
were insensitive to the vinyl ketones.[8] In order to
examine the vinyl ketone-sensitive enzymes involved in
aldehyde catabolism, [11,12-3H]-(Z)-1,11-hexadecadien-3-
one was synthesized.[9] Using the [3H]-vinyl ketone to
detect PAGE and SDS-PAGE separated proteins by
fluorography, aldehyde dehydrogenase (ALDH) isozymes were
found in antennae, head, and traces in leg tissues.[10]
Covalent modification was shown to be associated with
irreversible Michael addition of an active site cysteine
residue to the reactive enone system.

Epoxide Binding and Hydration: *Lymantria dispar*

Both enantiomers of the gypsy moth pheromone
disparlure were synthesized with specific activity
58 Ci/mmol and enantiomeric purity >95%.[11] Both the
attractive (+) enantiomer and behavior-inhibiting
(-)-enantiomer bind to one of a pair of homologous 14 kD
antennal PBPs, as detected by non-denaturing PAGE combined
with "native fluorography" using salicylate enhancement.[12]
For active site mapping, a photoaffinity label analog has
been prepared based on the aryl trifluormethyl diazirine
(S.McG. Graham, unpublished results).

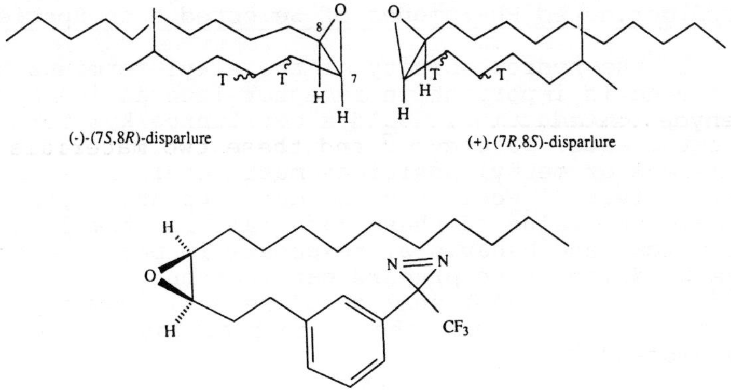

(-)-(7S,8R)-disparlure

(+)-(7R,8S)-disparlure

Figure 4   [³H]-Disparlure enantiomers and photolabel

Using radio-TLC, we determined that the epoxide is opened by epoxide hydrases (EH) distributed over the entire body, with particular concentrations in male antennae. Capillary GC of bis(TMS), bisacetate, and n-butylboronate derivatives of the diol product conclusively show that only the *threo* diol is produced by antennal EH after 24 hr at concentrations of 0.1 to 10 μM. Moreover, the enantiomeric compositions of the diols produced by epoxide opening of (+) and (-) disparlure were determined by chiral capillary GC of the bis(trifluoro-acetate) derivatives.[13] To our surprise, both the *(7R,8S)* and *(7S,8R)*-epoxide enantiomers produce the same *(7R,8R)* diol (Figure 5 shows result for (+)-disparlure). For a single enzyme to give this result, the enzyme must be able to accept either the branched hydrocarbon chain or the n-decyl chain in each of two hydrophobic pockets.[13]

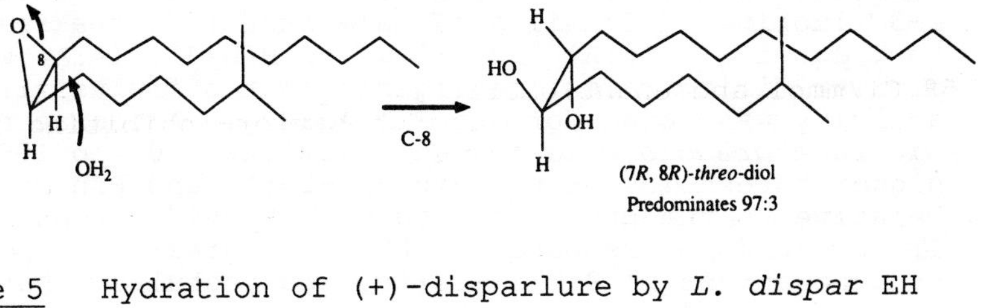

(7R, 8R)-*threo*-diol
Predominates 97:3

Figure 5   Hydration of (+)-disparlure by *L. dispar* EH

Polyfluorinated Pheromones of Selected Moth Species

   If the hydrophobicity of the alkyl terminus of a
pheromone is important in receptor interaction and thus
sensory transduction, would a perfluoroalkyl terminus
confer greater potency or reduced potency? Which
methylene or methyl positions must retain hydrogens for
full activity? Would the increased volatility hinder or
enhance the value of these materials in stimulating
electrical and behavioral responses in moths? These
questions led us to prepare perfluorobutyl (Pfb) and
perfluorohexyl (Pfh) analogs of several common pheromone
constituents[14] to test the concept of "non-stick
pheromones".[15]

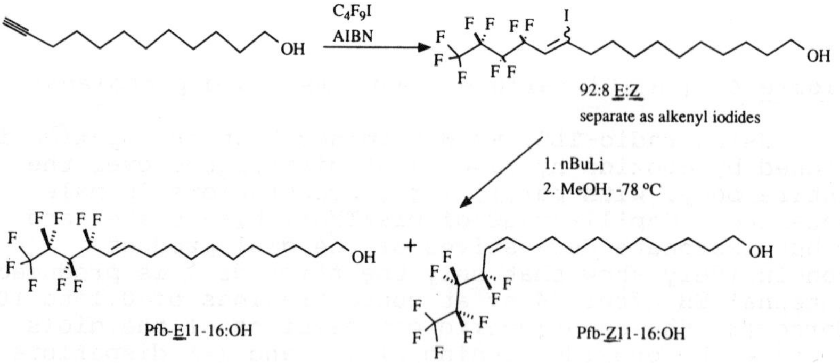

Figure 6   Synthesis of perfluorobutyl analog of Z11-16:OH

   Synthesis of pure *(Z)* isomers of Pfb and Pfh
compounds was accomplished as illustrated in Figure 6 for
Pfb-Z11-16:OH. These perfluoroalkyl analogs were
substantially more volatile than the corresponding
nonfluorinated parent compound, showing an average
decrease in Kovats retention of 1 methylene equivalent per
2-3 fluorines. Sensilla of male *Heliothis zea* responded
to 1 μg of Z11-16:Al and 250 μg of Pfb-Z11-16:Al with
similar firing rates; negligible post-stimulus firing
activity was found for the Pfb analog. Antennae of
Diatraea grandiosella showed a similar 100- to 1000-fold
higher threshold for the Pfb-Z11-16:Al and Pfh-Z9-16:Al
relative to the parent compounds. Specific sensilla of
Trichoplusia ni respond at 1000-fold greater
concentrations of Pfb Z7-12:Ac. We postulate that the
"slipperiness" or apparent fast off-rate, can be
attributed to reduced interaction between the hydrophobic

protein binding site and the more rigid and more polar perfluoroalkyl moiety.

However, not all fluorinations are detrimental. Preliminary evidence (M. Bengtsson, S. Rauscher, H. Arn, W-C. Sun, and G. Prestwich, unpublished results) suggests that an allylic difluoro analog of Z9-12:Ac retains high biological activity for selected tortricids, even though the perfluoroethyl analogs show very low activity as pheromone mimics.

3 JUVENILE HORMONE BIOCHEMISTRY

The primary unsolved problem in insect morphogenesis is how juvenile hormone (JH) and 20-hydroxyecdysone interact with putative receptor proteins and regulatory sequences of DNA to control new protein synthesis. While many extracellular JH binding proteins are documented through competitive binding assays[16] and more recently by photoaffinity labeling,[17,18] little is known of the molecular interactions of JH with intracellular (cytosolic and nuclear) proteins in target tissues. However, advances in molecular biology and radioligand design now offer ways to characterize these molecular sites of action more completely. The chemistry of high specific activity tritium- and iodine-labeled JH homologs and JH analogs are summarized, and selected recent uses are highlighted.[19] Of particular importance is the result that a methoprene analog and JH I bind to different binding sites in nuclear proteins of larval *Manduca* epidermis.[20] Does this mean there is a distinct class of methoprene receptors?

JH Homologs

The synthesis of tritium-labeled JH homologs of high enantiomeric purity was necessary for unambiguous measurement of binding affinities and rates of enzymatic processing. Thus, we prepared natural (10R,11S)-JH I and JH II at high specific activity (58 Ci/mmol) and enantiomeric purity (>95%) by tritiation of a vinyl oxirane obtained via an asymmetric epoxidation.[21] This synthesis is now being used by New England Nuclear for commercial production of labeled JH I and by Sigma for unlabeled JH I. A high specific activity (11-17 Ci/mmol), chiral JH III was prepared by a reductive deoxygenation,[22] by a different route than that employed for labeled racemic JH III by Zoecon and NEN.[23] Binding specificity has been shown in four caterpillars using these pure labeled enantiomers.[24,25] Purified JH esterase from two heliothine caterpillars is also enantioselective.[26]

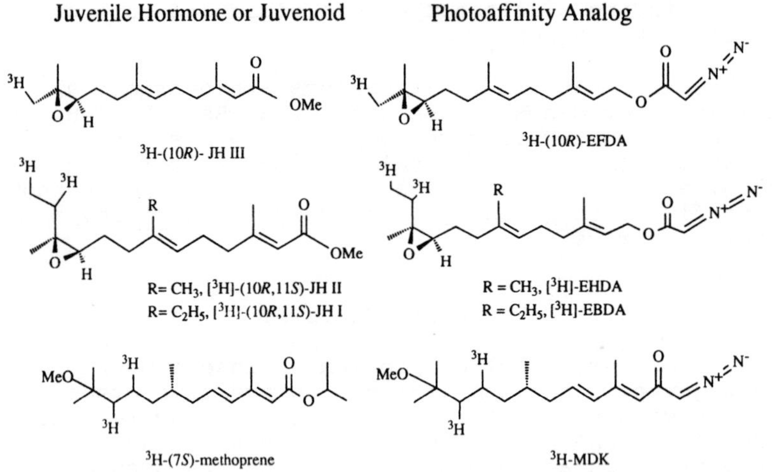

Juvenile Hormone or Juvenoid   Photoaffinity Analog

^{3}H-(10*R*)- JH III

^{3}H-(10*R*)-EFDA

R= CH$_3$, [^{3}H]-(10*R*,11*S*)-JH II
R= C$_2$H$_5$, [^{3}H]-(10*R*,11*S*)-JH I

R = CH$_3$, [^{3}H]-EHDA
R = C$_2$H$_5$, [^{3}H]-EBDA

^{3}H-(7*S*)-methoprene

^{3}H-MDK

Figure 7   Radiolabeled natural JHs, juvenoids, and photolabels

JH and Juvenoid Photoaffinity Labels

Our first experiments with [^{3}H]-EFDA, a JH III analog, showed the utility of this material for cockroach[27] and caterpillar[28] JHBPs. Subsequently, we have prepared diazoacetate analogs of JH II ([^{3}H]-EHDA) and JH I ([^{3}H]-EBDA),[29] which show up to twenty-fold higher affinity for labeling caterpillar JHBPs.[30] Moreover, these analogs are physiologically-active in the black *Manduca* assay and can be used to covalently modify nuclear proteins in fat body and epidermis of *Manduca* larvae.[31] All three diazoacetates are now most efficiently prepared in optically-active form in two steps from the parent [^{3}H]-JH homolog.[32]

New synthetic schemes were developed to allow access to high specific activity [^{3}H]-(7*S*)-hydroprene[33] and [^{3}H]-(7*S*)-methoprene.[34] The key to the strategy was the tritiation of an isolated alkene with the dienoate moiety protected as an iron tricarbonyl adduct. [^{3}H]-Methoprene has been employed to produce [^{3}H]-methoprene diazomethyl ketone ([^{3}H]-MDK), an important radioligand for nuclear receptor characterizations.[31]

Phenoxyphenyl ether IGRs

The phenoxyphenyl ether IGRs are among the most potent non-sesquiterpenoid juvenoids. We used traditional reductive tritiodebromination to prepare selectively monotritiated [3H]-fenoxycarb and [3H]-S-31183.[35] Binding assays, penetration assays, and preparation of photoaffinity labeled IGRs are topics of ongoing research.

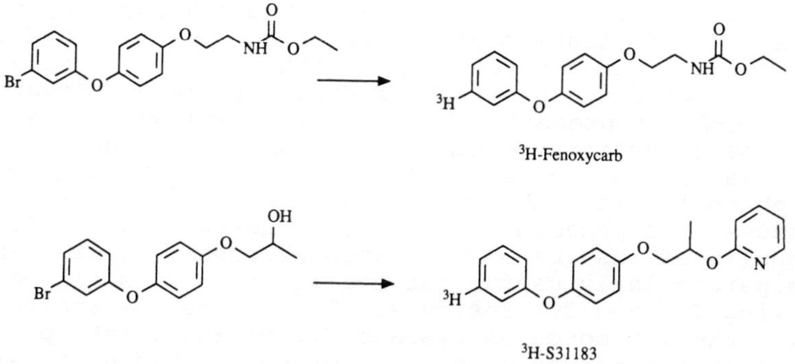

3H-Fenoxycarb

3H-S31183

Figure 8 Preparation of [3H]-IGRs

Iodinated Juvenoids

Two radioiodinated species have been prepared to provide high specific activity gamma emitters for receptor characterization. First, we synthesized [125I]-IVMA,[36] a methoprene analog, to search for cellular binding proteins in *Manduca* epidermis; surprisingly, high affinity binding (3 nM) was observed in nuclei but JH I did not compete for this binding site![37]

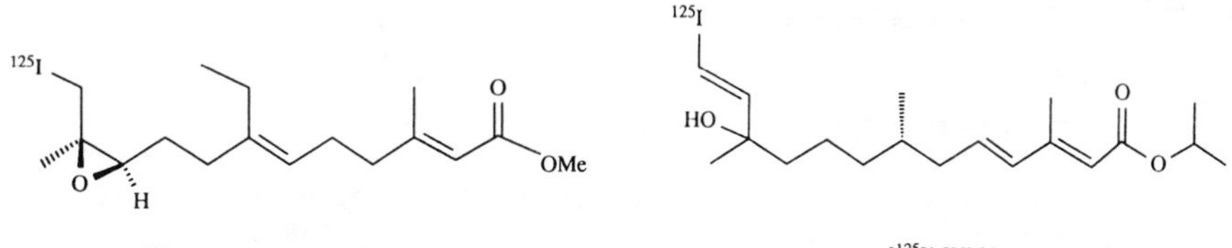

[125I]-(10*R*,11*S*)-12-iodo-JH I

[125I]-IVMA

Figure 9 Radioiodinated JH I and methoprene analogs

Next, a physiologically-active JH I analog was prepared by isosteric substitution of iodine for methyl. The [125I]-12-iodo-JH I showed high affinity for hemolymph

binding proteins[38] and led to the development of a
sensitive technique for peptide sequencing from
electroblotted, radioligand-bound native proteins.[39]

<u>Peptide hormones</u>

As yet unexplored are receptor sites for insect
peptide hormones involved in developmental processes, such
as eclosion hormone,[40] allatotropin,[41] and PTTH.[42] Analogs
bearing radiolabels and photoattachable groups would permit
covalent modification of target tissue proteins mediating
hormone action. In connection with work on crustacean
peptide hormone action, an azidosalicylamide derivative of
Glu[1]-CC-2 was prepared, the unblocked analog of the
cardioacceleratory peptide of the American cockroach.[43]
This was radiolabeled and [^{125}I]-asa-Glu[1]-CC-2 was employed
to photoaffinity label membrane receptor proteins in
tissues of a penaeid shrimp (G.D. Prestwich and E.S. Chang,
unpublished results). Iodinated azidosalicylamide analogs
are particularly useful for modification of small peptides
lacking Tyr and Lys residues. An alternative approach
using the introduction Lys and Tyr residues into peptide by
total synthesis was used to generate antibodies to AKH;[44]
this approach would also be amenable to introduction of
photolabels.

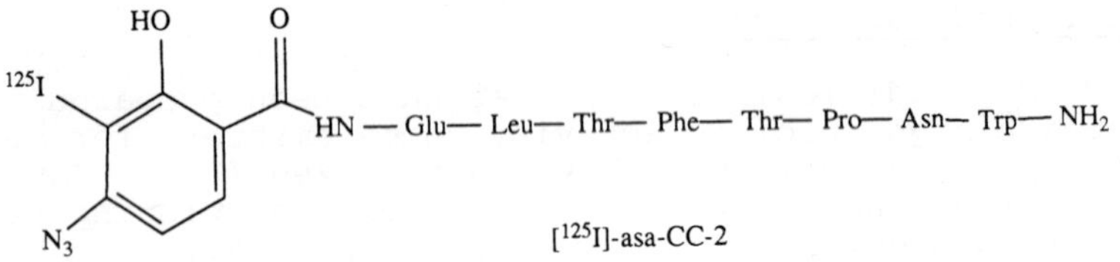

[^{125}I]-asa-CC-2

<u>Figure 10</u> Photoaffinity label for invertebrate peptide
receptors

5 REFERENCES

1. R.G. Vogt, 'Pheromone Biochemistry', (G.D. Prestwich and
    G.J. Blomquist, eds.) Academic Press, NY, 1987.
2. R.G. Vogt, G.D. Prestwich, and L.M. Riddiford, <u>J. Biol. Chem.</u>,
    1988, <u>263</u>, 3952.
3. B. Latli and G.D. Prestwich, <u>J. Org. Chem.</u>, 1988, <u>53</u>, 4603.
4. P.E.A. Teal and J.H. Tumlinson, <u>J. Chem. Ecol.</u>, 1986, <u>12</u>, 353.
5. Y.-S. Ding and G.D. Prestwich, <u>J. Chem. Ecol.</u>, 1986, <u>12</u>, 411.
6. Y.-S. Ding and G.D. Prestwich, <u>J. Chem. Ecol.</u>, 1988, <u>14</u>, 2033.

7. G.D. Prestwich, M.L. Tasayco J., B. Latli, M. Handley, L. Streinz, and S.McG. Graham, Experientia, 1989, 45, 263.
8. M.L. Tasayco J. and G.D. Prestwich, J. Biol. Chem., submitted (1989a).
9. G.D. Prestwich, Pure Appl. Chem., 1989, 61, 551.
10. M.L. Tasayco J. and G.D. Prestwich, J. Biol. Chem., submitted (1989b).
11. G.D. Prestwich, S.McG. Graham, J.-W. Kuo, and R.G. Vogt, J. Am. Chem. Soc., 1989, 111, 636.
12. R.G. Vogt, A.C. Koehne, J.T. Dubnau, and G.D. Prestwich, J. Neurosci., in press (1989).
13. G.D. Prestwich, S.McG. Graham, and W.A. Koenig, J. Chem. Soc. ,Chem. Commun., 1989, 575.
14. G.D. Prestwich, W.-C. Sun, M.S. Mayer, and J.C. Dickens, J. Chem. Ecol., submitted (1989).
15. G.D. Prestwich and J.A. Pickett, in preparation.
16. W.G. Goodman and E.S. Chang, 'Comprehensive Insect Biochemistry, Physiology, and Pharmacology', Pergamon Press, Oxford, 1985.
17. J.K. Koeppe and G.E. Kovalick, Biochem. Actions Hormones, 1986, 13, 265.
18. G.D. Prestwich, J.K. Koeppe, G.E. Kovalick, J.J. Brown, E.S. Chang, and A.K. Singh, Methods in Enzymology (Steroids and Isoprenoids), 1985, 111, 509.
19. G.D. Prestwich, Science, 1987, 237, 999.
20. E.O. Osir and L.M. Riddiford, J. Biol. Chem., 1988, 263, 13812.
21. G.D. Prestwich and C. Wawrzeńczyk, Proc. Natl. Acad. Sci. USA, 1985, 82, 5290.
22. W.-s. Eng and G.D. Prestwich, J. Labelled Cmpnd. Radiopharm., 1988, 25, 627.
23. F.C. Baker and D.A. Schooley, J. Labelled Cmpnd. Radiopharm., 1986, 23, 533.
24. G.D. Prestwich, S. Robles, C. Wawrzeńczyk, and A. Bühler, Insect Biochem., 1987, 17, 551.
25. J.R. Wiśniewski, C. Wawrzeńczyk, G.D. Prestwich, and M. Kochman, Insect Biochem., 1988, 18, 29.
26. Y.A.I. Abdel-Aal, T.N. Hanzlik, B.D. Hammock, L. Harshman, and G.D. Prestwich, Comp. Biochem. Physiol., 1988, 90B, 117.
27. J.K. Koeppe, G.E. Kovalick, and G.D. Prestwich, J. Biol. Chem., 1984, 259, 3219.
28. J.K. Koeppe, G.D. Prestwich, J.J. Brown, W.G. Goodman, G.E. Kovalick, T. Briers, M.D. Pak, and L.I. Gilbert, Biochemistry, 1984, 23, 6674.
29. W.-s. Eng, PhD Thesis, State University of New York, Stony Brook, New York, 1987.
30. G.D. Prestwich, M.F. Boehm, W.-s. Eng, P. Kulcsár, N. Maldonado, S. Robles, U. Sinha, and C. Wawrzeńczyk, 'Endocrinological Frontiers of Physiological Insect Ecology',Wrocław Technical University Press, Wrocław, Poland, 1988.

31.   S.R. Palli, E. Osir, M.F. Boehm, W.-s. Eng, I. Ujváry,
       M. Edwards, P. Kulcsár, G.D. Prestwich, and L.M. Riddiford,
       Proc. Natl. Acad. Sci. USA, in press (1989).
32.   I. Ujváry and G. D. Prestwich, in preparation for J. Labelled
       Cmpnd. Radiopharm., submitted (1989).
33.   M.F. Boehm and G.D. Prestwich, J. Org. Chem., 1987, 52, 1349.
34.   M.F. Boehm and G.D. Prestwich, J. Labelled Cmpnd. Radiopharm.,
       1988, 25, 653.
35.   M.F. Boehm and G.D. Prestwich, J. Labelled Cmpnd. Radiopharm.,
       1988, 25, 1007.
36.   M.F. Boehm and G.D. Prestwich, J. Org. Chem., 1986, 51, 5447.
37.   L.M. Riddiford, E.O. Osir, C.M. Fittinghoff, and J.M. Green,
       Insect Biochem., 1987, 17, 1039.
38.   G.D. Prestwich, W.-s. Eng, S. Robles, R.G. Vogt, J. Wisñiewski,
       and C. Wawrzeñczyk, J. Biol. Chem., 1988, 263, 1398.
39.   P. Kulcsár and G.D. Prestwich, FEBS Lett., 1988, 228, 49.
40.   T. Marti, K. Takio, K.A. Walsh, G. Terzi, and J.W. Truman,
       FEBS Lett., 1987, 219, 415.
41.   Kataoka, H., A. Toschi, J.P. Li, R.L. Carney, D.A. Schooley, and
       S.J. Kramer, Science, 1989, 243, 1481.
42.   H. Nagasawa, H. Kataoka, A., Isogai, S. Tamura, A. Suzuki,
       A. Mizogushi, Y. Fujiwara, A. Suzuki, S.Y. Takahashi,
       H. Ishizaki, Proc. Natl. Acad. Sci. USA, 1986, 83, 5840.
43.   R.M. Scarborough, G.C. Jamieson, F. Kalisch, S.J. Kramer,
       G.A. McEnroe, C.A. Miller, and D.A. Schooley, Proc. Natl. Acad.
       Sci. USA, 1984, 81, 5575.
44.   S. Hekimi and M. O'Shea, Insect Biochem., 1989, 19, 79.

Three-dimensional Mapping of Insect Taste Receptor Sites as an Aid to Novel Antifeedant Development

J. L. Frazier and P. Y-S.Lam

AGRICULTURAL PRODUCTS AND MEDICAL PRODUCTS DEPARTMENTS, E. I. DUPONT DE NEMOURS, WILMINGTON, DELAWARE 19880-0402, USA

1 INTRODUCTION

Insects possess highly-adapted behaviors for mate recognition, feeding, and egg laying that are elicited in the appropriate context by specific chemical sign stimuli. These
signals are detected by numerous types of chemosensory
cells that are grouped within sensilla located primarily
on the mouthparts, antennae, tarsi, and ovipositors. Chemosensory cells on the mouthparts of caterpillars that respond to taste stimuli are among the best studied, and were
selected for a biorational design of antifeedants.

The styloconica of the tobacco hornworm, like those
of many caterpillars contain cells responsive to plant
sugars [1,2,3]. Information from these cells is used by the
caterpillar in the decision to feed. Although the exact
sensory codes are yet to be understood, some antifeedants
appear to disrupt the messages from sugar-sensitive cells
and reduce feeding [4,5,6]. Mapping the binding requirements
for the receptive sites of these sugar-sensitive cells and
using this information to design and synthesize inhibitors
was the approach we selected.

2 QUANTIFYING CHEMOSENSORY CELL ACTIVITY

The responses of chemosensory cells are recorded by applying a candidate stimulus solution in a recording electrode
pipette to the tip of a sensillum and monitoring the resulting action potentials. Since the responses of all the
excited chemosensory cells are recorded together, numerous
artifacts resulting from partially summated action potentials are often produced (Figure 1). Separating the re-

sponses of individual taste cells from each other and from
the attendant artifacts has been a major impediment in the
past, but the recent application of computer aided analysis
has greatly facilitated this task [7]. A template matching

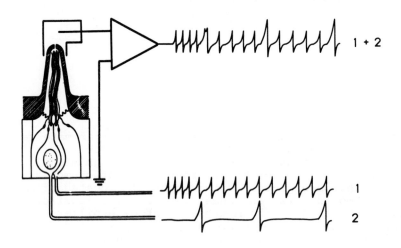

<u>Figure 1</u> Tip-recording technique produces multiunit action
potential records that require separation into single units

technique is used to identify the action potentials produced by each cell (Figure 2). This figure shows a computer reconstructed train of 64 action potentials for a one
second stimulation in the first trace. The template for
the glucose-sensitive cell is shown to the left of the
second trace, with all spikes that match the template within 20% variation indicated by a vertical line under each
spike. In this trial, 39 spikes were produced by the
glucose-sensitive cell. The template for the salt-sensitive cell is shown to the left of the third trace, with all
spikes that match it indicated. In this case, 20 spikes
were produced by the salt-sensitive cell. Noise pulses
that did not match either template are indicated in the
fourth trace labelled G. Action potentials that did not
match either template are shown in the fifth trace labelled
U. In this way the responses of each cell type are identified and can be quantitated and related to the concentration of a stimulus compound or to the relative potencies
of individual compounds. One can thus approach QSAR studies for individual taste cells within a reasonable time-

frame, that previously was not possible.

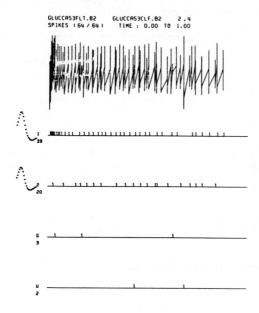

<u>Figure 2</u> Template matching separates responses of the glucose-sensitive taste cell from that of a salt-sensitive cell

3 MAPPING THE GLUCOSE RECEPTOR SITES

Our approach combines that used for characterizing the sucrose receptor sites of the flesh fly [8] and the gerbil [9] and the use of fluorinated carbohydrates as probes for enzyme sites [10]. Details of this study are reported elsewhere [11], but the rationale and some highlights are given here.

If the deoxy-glucose at each carbon position produces stimulation less than glucose, then hydrogen bonding is deemed to be important at that position. Since fluorine is the closest substitute for oxygen as a hydrogen bond acceptor in electronic and steric terms, then comparison of the stimulating effectiveness of fluorine substitutions at each carbon position with the deoxy- analog can reveal the direction of hydrogen bonding. If the fluoro-glucose has equal stimulating effectiveness with the corresponding deoxy-glucose, then the hydroxyl group is serving as a hydrogen bond donor to the receptor.

Table 1 gives a summary of the stimulating effectiveness of the various fluoro- and deoxy-glucoses used to determine the hydrogen bonding pattern for the medial glucose-sensitive cell of the tobacco hornworm caterpillar. For the C-1 position, 1,5-anhydroglucitol is a weak stimulus relative to glucose, but both the 1-alpha and 1-beta fluoro-glucoses are equal to glucose, indicating that the oxygen is serving as a hydrogen bond acceptor from the receptor. For the C-2 position, 2-deoxy-glucose and the 2-fluoro-glucose are both weak stimuli, indicating that hydrogen bonding is important, and that the stimulus is acting as a hydrogen bond donor to the receptor. It can be seen from these data, that the C-1 and C-3 positions are involved as hydrogen bond acceptors and donors whereas the other positions mapped are involved as hydrogen bond donors. In terms of hydrogen bond donating ability, the

Table 1 Determination of Binding Atom at each C Position by Comparison of Stimulating Effectiveness of Deoxy- and Fluoroglucoses for the Medial Cell

Compound	Stimulation As% Control ±SEM	Binding Atom	Hydrogen Bonding to Receptor
α-D Glucose	100± 9	--	
1,5-Anhydroglucitol	25± 8	--	
1-α-Fluoro-Glucose	93±11	O	Acceptor
1-β-Fluoro-Glucose	113±13	O	
2-Deoxy-Glucose	16± 2		
2-Fluoro-Glucose	12± 2	H	Donor
3-Deoxy-Glucose	24± 4		Acceptor
3-Fluoro-Glucose	72± 9	O,H	And Donor
4-Deoxy-Glucose	25± 2		
4-Fluoro-Glucose	28± 7	H	Donor
6-Deoxy-Glucose	33± 6		
6-Fluoro-Glucose	43±13	H	Donor

N = 6-8 Cells/Compound

following order is observed: C-2>C-4>C-6>C-3>C-1.

The rationale for determining the three-dimensional aspects of the receptor hydrogen bonding relative to the plane of the stimulus glucose combined the use of epimers at each carbon position relative to glucose. The assumptions are that stimulating effectiveness increases with stronger hydrogen bonding, and that closer proximity yields stronger hydrogen bonding. If the epimeric hydroxy at a certain position contributes to binding, the receptor site is located between the hydroxy positions of the epimer and glucose. In this mode, the receptor site will be situated approximately in the plane of the glucose ring. On the other hand, if the epimeric hydroxy at a certain position does not contribute to binding, the receptor site will be located on the opposite side of the epimeric position. In this mode, the receptor site will be either above or below the plane of the glucose ring.

It can be seen from Table 2, that the position of the receptor hydrogen bond acceptor is above the plane for the C-3 position and on the plane for the C-2 and C-4 positions. Likewise, the receptor hydrogen bond donor posi-

Table 2 Three Dimensional Position of Hydrogen Bonding at each C Position for the Medial Glucose Receptor Site

Glucose Epimer/ Fluoro Analogues	Epimeric At	Binding Atom	Percent Stimulation (+SEM)	Position of Receptor H-Bond Acceptor	Position of Receptor H-Bond Donor
α-D-Glucose			100+ 9	---	
1-α-Fluoroglucose 2-β-Fluoroglucose	C-1	0	93+11 113+13		On the Plane
D-Mannose 2-Fluoromannose	C-2	H	74+ 6 6+ 3	On the Plane	---
D-Allose 3-Fluoroallose	C-3	O,H	67+ 5 83+ 4	Above Plane	On the Plane
D-Galactose 4-Fluorogalactose	C-4	H	83+ 6 0	On the Plane	---
L-Idose	C-6	H	63+ 6	?	---

N = 6-8 Cells/Compound

tions are on the plane for the C-1 and C-3 positions.

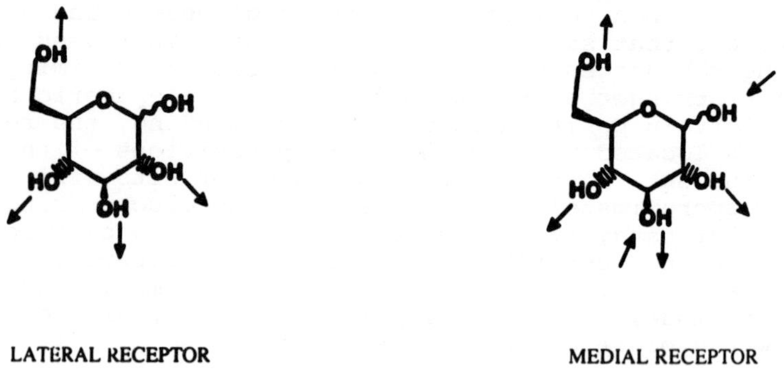

LATERAL RECEPTOR   MEDIAL RECEPTOR

Figure 3 Comparison of the requirements for binding between
the lateral and medial glucose receptor sites of the tobacco hornworm styloconic cells. Arrows indicate the direction of bonding.

    A comparison of the binding site requirements for the
medial and lateral glucose-sensitive cells is shown in
Figure 3. The lateral site is less specific than the medial site, with hydrogen bond donation from the stimulus to
the receptor at the C-2, C-3, C-4, and C-6 positions. In
contrast, the medial site involves hydrogen bond donation
at the same positions as the lateral with the addition of
the hydroxyl oxygens acting as hydrogen bond acceptors at
the C-1 and C-3 positions. The medial site thus has additional and more specific binding requirements than the
lateral site. The specificities of these caterpillar sites
differ from those reported for the fly pyranose subsite[8].

    4 SYNTHESIS AND EVALUATION OF GLUCOSE SITE INHIBITORS

Based on the above binding site models and the knowledge
that the C-1, C-3, and C-6 positions are not as critical
for binding to the receptor as C-2 and C-4, a series of
compounds with electrophiles were synthesized as active-
site directed irreversible inhibitors [12,13]. The electro-

philes were either alkylating agents or Michael acceptors exhibiting a wide range of activities. Previous studies had indicated that a functionally important thiol group was present in the vicinity of the receptor[6]. The compounds were tested for their antifeedant activity on several species of insects, with selected compounds analyzed further for their effects on the glucose-sensitive taste cells of the tobacco hornworm.

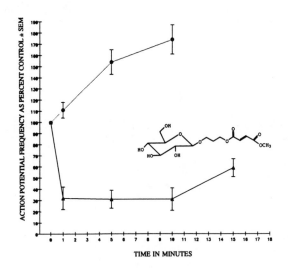

TIME IN MINUTES

<u>Figure 4</u> Blocking of a glucose-sensitive taste cell by a synthesized candidate antifeedant

The C-1 fumarate derivative of glucose was shown to produce total reduction of glucose stimulated feeding in the tobacco hornworm, with lesser effects on host plant feeding for this as well as other species (Table 3). Studies of the action of this compound at the taste cell level indicated that it partially inhibited the response of the glucose-sensitive cell (Figure 4). Following a two minute treatment of the cell with 0.1 m glucose, the response recovers rapidly and often shows a post rebound effect to higher than control levels. The same cell treated with equimolar fumarate for two minutes shows a 70% reduction in response that persists for more than 15 minutes before recovering. Other glucose derivatives exhibited similar, but shorter duration inhibitory effects.

Table 3 Reduction in Feeding by Candidate Glucose
Inhibitors on Several Caterpillar Species

Compound	Percent Reduction in Feeding by 1000 PPM on Leaf Disks	Species
	77	Heliothis virescens
	15	Spodoptera frugiperda
	33	Heliothis virescens
	63	Spodoptera frugiperda
	58	Ostrinia nubilalis

The styloconic glucose-sensitive taste cells of the tobacco hornworm have served as models for their receptor site mapping and for the design of candidate inhibitors. These two sites exhibit functional differences that are reflected in the efficacy of various compounds for stimulating and inhibiting them. These same compounds also exhibit various low potency activities on other caterpillar species, for which we know nothing about their receptor site geometries. Although these rationally designed compounds reduced taste cell inputs with concomitant reduced feeding, perhaps the evolutionary arena of insect taste and plant distaste is even more complex than previously imagined.

ACKNOWLEDGEMENTS

The authors gratefully acknowledge the expert technical assistance of Bob Croes, Lisa Chapaitis, and Dan Cordova and Leslie Savage for typing the manuscript.

REFERENCES

1. L.M. Schoonhoven, "Perspectives in Chemoreception and Behavior" E. Bernays and R. Chapman (Eds) Springer-Verlag, N.Y., 1987, p.69.
2. W.M. Blaney and M.S.J. Simmonds, Entomol. Exp. Appl., 1988, 49, 111.
3. J.L. Frazier, "Molecular Aspects of Insect-Plant Associations", L.B. Brattsten and S. Ahmad (Eds) Plenum, N.Y., 1986, p.1.
4. W.C. Ma, Physiol. Ent., 1977, 2, 199.
5. W.M. Blaney and M.S.J. Simmonds, Entomol. Exp. Appl., 1984, 36, 209.
6. J.L. Frazier and P.Y-S. Lam, Chemical Senses, 1986, 11, 600.
7. J.L. Frazier and F.E. Hanson, "Insect/Plant Interactions", J.R. Miller and T.A. Miller (Eds), Springer-Verlag, N.Y., 1986, p. 385.
8. I. Shimada, Chemical Senses, 1987, 12, 235.
9. W. Jakinovich,Jr. "Taste, Olfaction, and the CNS", D.W. Pfall (Ed.), Rockefeller U. Press, N.Y., 1985, 65.
10. S.G. Withers, I.P. Street and M.D. Percival, A.C.S. Symp., 374, 1988, p. 59.
11. P.Y. Lam and J.L. Frazier, Chemical Senses, 1989, submitted.
12. B. Classon, P.J. Garegg and B. Sameulsson, Acta Chemica Scan. 38, 1984, 419.
13. Y.Oikawa, T.Yoshioka, and O.Yonemitsu, Tetra.Lett. 1982, 23, 885.

Use of Transition State Theory in the Design of Chemical and Molecular Agents for Insect Control

B. D. Hammock,*† A. Székács, * T. Hanzlik, * S. Maeda, * M. Philpott,* B. Bonning, *, and R. Possee †

DEPARTMENTS OF ENTOMOLOGY AND ENVIRONMENTAL TOXICOLOGY, UNIVERSITY OF CALIFORNIA, DAVIS, CALIFORNIA 95616, USA*
NERC INSTITUTE OF VIROLOGY AND ENVIRONMENTAL MICROBIOLOGY, MANSFIELD ROAD, OXFORD OX1 3SR, UK†

1. INTRODUCTION

This manuscript has two main purposes. The first and minor purpose is to argue for an integration of chemical and molecular approaches in agricultural chemistry as a mechanism to increase the number of useful leads in each discipline. The second and major purpose of the manuscript is to illustrate this integration on a small scale with our work on disruption of the insect endocrine system. This work involves a combination of a number of biological, biochemical, and molecular approaches, but it relies heavily on the use of transition state theory to synthesize inhibitors and affinity purification systems for a key regulatory enzyme. Using these reagents we present very encouraging evidence for the development of a genetically engineered viral insecticide.

As one would expect from both the history and the name of agricultural chemical companies, most companies are directed by chemists and have a largely chemical approach to the discovery of new materials for insect control. Although this approach has been very successful, the limitation of screening in the discovery of commercially useful materials has been discussed by many workers and is well illustrated by the decreased number of companies actively involved in research and discovery programs.[1,2] There are various new approaches to increase the proportion of successes in the field, and an investment in recombinant DNA technology certainly represents an approach taken by many industries.

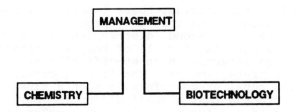

Figure 1   There often is a lack of
integration of chemical and biochemical
approaches to the discovery of pest
control agents.

Although most companies have invested in such new
technology, cynically one might feel that this
investment was more for appearance or insurance than for
any great faith in the short term success of recombinant
DNA technology. In general, the molecular and chemical
approaches to agricultural problems have progressed
along divergent lines and are isolated by administrative
and certainly intellectual barriers. This intellectual
isolation may be reciprocated as molecular biologists
see limited success following a larger investment in
classical chemical approaches to pest control. Thus,
molecular biologists restrict themselves largely to
'biochemical' leads, while chemists take their leads
from a variety of sources such as the natural product
literature, enzymes and receptor studies, and primarily
from screening. It is this last approach that
represents the major investment of resources, yet often
the leads resulting from biological activity discovered
in a screening process are lost to the molecular
biologists. It seems clear that if the intellectual and
administrative barriers can be eliminated between these
fields both approaches will prosper. Clearly a major
goal must be the devising of procedures enabling the
chemist and biologist to exploit the tremendous data
base on compounds which show desirable biological
activities, but do not yield compounds that can be
commercialized.

2.   BIOLOGICAL TARGET

In our work we have targeted regulation of development
in moths in the family Noctuiidae for several reasons.
First, if we were successful in finding a family or even

genus specific insecticide, these insects are such
serious worldwide pests that they probably would support
development costs. Second, discoveries here may lead to
broader range material for all of the Lepidoptera or all
holometabolous insects. More specifically we have
targeted the transition from the feeding caterpillar
stage to the pupal stage. This is such a finely
controlled process that a small disruption may lead to a
major effect on the insect. Also, such metamorphosis is
unique to insects so that there is the hope of low
toxicity to non-insect species.

Simplistically, the insect molts when its tissues
are exposed to a water soluble steroid termed 20-
hydroxyecdysone. If the insect encounters ecdysone in
the presence of a terpenoid hormone termed juvenile
hormone (JH), then the insect undergoes an isometric
molt. That is, it increases in size, but there is no
change in form. If the tissues encounter ecdysone in
the presence of low JH then an anisometric molt results
in the formation of a more mature stage which in normal
circumstances is a pupa.[3]

Juvenile hormone has been the target of many
efforts to develop insect control agents, and several
products are on or near the market based on their
properties as JH agonists or partial agonists. These
materials are very successful in the control of pests
where damage is done by the adults. However, they act
by extending the larval stage and potentially increasing
damage to crops. Thus, for many pests of field and row
crops, an anti-JH has not yet reached commercial
development. Like many other laboratories, we have
attempted to reduce the titer of JH by the development
of inhibitors of biosynthesis. However, there is an
alternative approach based upon an understanding of
insect development. That approach is to increase the
degradation of JH in the insect.

At least in the lepidopterous insect studies, the
JH titer decreases in preparation for pupation as a
result of a reduction in biosynthesis by the corpora
allata and the production of a highly aggressive enzyme
known as juvenile hormone esterase (JHE). This enzyme
hydrolyzes the stable, conjugated methyl ester of
juvenile hormone to the corresponding JH acid which is
biologically inactive in most assays. The enzyme is
produced several times during development and its
regulation appears complex. For the purpose of this
argument, key appearances include a small burst of

activity at ecdysis where the resulting decrease in JH
may reduce feeding behavior and in the first few days of
the last larval stadium where the reduction in JH titer
reduces feeding. The reduction in the JH titer permits
the tissues to reprogram for pupal development when they
encounter ecdysone. Because of its importance in
development, we have studied JHE for several years. A
key approach has been to make inhibitors of the enzyme
based on an appreciation of its catalytic mechanism.[4]

3. INHIBITION OF JHE WITH TRANSITION STATE MIMICS

Classical Inhibitors of JHE

It was noted early that JHE was very resistant to a
variety of classical esterase inhibitors including
eserine, carbaryl and DFP. However the enzyme could be
inhibited by compounds such as paraoxon. To study its
role in vivo it was important to have nontoxic, but
potent inhibitors of this enzyme. A screen of a variety
of compounds gave the surprising result that
methamidophos was a far more potent inhibitor of JHE
than of AChE (its theoretical target). Further work
with the phosphoramidothiolates resulted in EPPAT (O-
ethyl, S-phenyl phosphoramidothioate) which was
sufficiently selective to inhibit JHE in vivo without
killing the insect. If over 80% of the JHE was
inhibited, the resulting insects maintained sufficient
JH to continue feeding and growing rather than pupating.
However, these compounds were quite toxic to the insects
and inhibited a variety of hemolymph esterases.[5]

Trifluoromethylketone Inhibitors of JHE

Based on the ideas of Pauling,[6,7] we synthesized
trifluoroketones which should mimic transient
intermediates along the reaction coordinate leading from
the initial Michaelis complex of JH and JHE through the
acyl enzyme to the product JH acid.[8] The most potent of
these compounds had linear free energy parameters
similar to that of JH itself. Although they proved to
be very selective and potent inhibitors in vitro, these
compounds were unable to inhibit sufficient esterase in
vivo to disrupt development. However, when a thioether
was placed beta to the carbonyl (right, below), the
compounds were 50-100 times more active than their
carbon analogs (left).

$C_{10}H_{21}C(O)CF_3$ $C_8H_{17}SCH_2C(O)CF_3$

Figure 2   Both the trifluoromethyl ketones
(left) and the substituted thiotrifluoropropanones (right) proved to be potent in vitro
inhibitors of JHE, but only the latter were
active in vivo.

Topical application of these thiotrifluoropropanones stabilized JH in vivo and kept the insects in
a feeding phase.[8,9] It is thought that the thioether
mimics the 2,3-olefin of JH and leads to better binding
of the inhibitor to the enzyme. However, X-ray
crystallography of the materials shows a surprisingly
strong hydrogen bond between the hydrated carbonyl and
the sulfur of the thioether.[10] If such a structure
exists on the enzyme it would stabilize the covalent
bond between the catalytic serine and the inhibitor and
may in part lead to the slow tight binding kinetics
described below. Similar results have been found
recently for trifluoroketones on nitrogen heterocycles
giving support to the above argument.

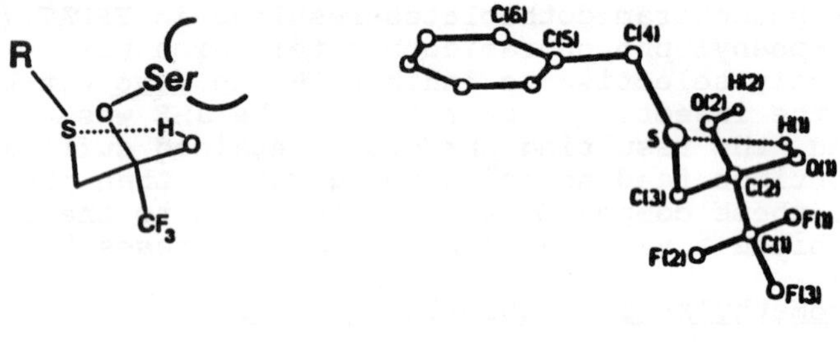

A   B

Figure 3   The tetrahedral structure of the thio-
trifluoropropanone-JHE adduct (A), possibly
stabilized by an intramolecular hydrogen bonding
interaction (shown as the dotted line). Based on
X-ray crystallography, similar strong hydrogen
bonds were found to stabilize the tetrahedral
hydrate of certain thiotrifluoropropanones (B).

Biological Implication of JHE Inhibition

This work is important biologically by
demonstrating for the first time that the titer of an

epithelial hormone is controlled in part by the stage
specific production of a catabolic enzyme. It also has
been important in demonstrating the use of transition
state theory as a paradigm in agricultural chemistry.
However, from an agricultural standpoint, the inhibitors
of JHE have never been seriously considered as
commercial compounds. Their dramatic biological
activity in generating very large and very hungry
larvae, is hardly desirable from the standpoint of a
farmer. Thus, these compounds might be considered as
models for the numerous leads which come from an
industrial screening program which are interesting, but
not of the correct activity or potent enough for
commercial exploitation. Our work in exploiting this
chemistry for insect control is outlined in section 4
below.

<u>Transition State Theory as a Paradigm for Discovery</u>

     The work on JHE demonstrates that transition state
theory, even when simplistically applied, can aid in the
development of highly potent inhibitors of an enzyme
with only a very few compounds being synthesized.[11-13]
More recent SAR approaches using many aliphatic,
aromatic and heterocyclic trifluoroketones continue to
support the concept that these compounds are mimicking
an intermediate of the reaction between JH and JHE.[14]
The most potent compounds do exhibit slow tight binding
kinetics which may help to explain their activity <u>in</u>
<u>vivo</u>. Such nonclassical kinetics probably are the
result of limitations of kinetic studies rather than a
magical change in molecular properties. However, they
support the long standing observation that materials
which bind covalently or in a poorly reversible fashion
to a biological target, are more likely to be a potent
<u>in</u> <u>vivo</u> than a potent <u>in</u> <u>vitro</u> agent that is rapidly
reversible.

     Although inhibition of JHE is not a viable
commercial target, polarized ketones in general and
trifluoroketones in particular may prove very useful as
inhibitors of a variety of biologically important
hydrolases. By varying the substituent on substituted
thio 1,1,1-trifluoropropan-2-ones, we have developed
potent inhibitors of a variety of esterases including
malathion and pyrethroid esterases, AChE, lipase, and
human brain neurotoxic esterase. By varying
substituents on polarized carbonyls further, it is
likely that potent inhibitors of a variety of serine

proteases and esterases acting on biologically important
molecules can be developed.

In designing inhibitors of catalytic processes,
biological activity may be found more quickly by
mimicking catalytic intermediates than by mimicking
substrates or products. Historically, one can consider
the binding of carbamate and phosphate insecticides to
AChE more dependent upon their similarity to a
tetrahedral intermediate than to their similarity to the
substrate acetylcholine.[12] More recent examples include
glyphosate and the sulfonyl ureas which probably act in
part as transition state mimics of their target enzymes.

4. AFFINITY PURIFICATION OF JHE

The richest source of JHE is the hemolymph of
caterpillars just before they begin to wander in the
last larval stadium. The enzyme presents a worst case
situation for purification by classical means, and it
may represent the situation with many of the most
promising targets for the development of biologically
active compounds. Either one must laboriously bleed
small animals or suffer a reduction in specific
activity. Although JHE is a very active and stable
enzyme, for which there is a highly sensitive and
specific assay, it is in relatively low abundance in the
presence of an extraordinarily high concentration of
other proteins.

Affinity chromatography thus represented a very
promising approach to the purification of JHE. By
inhibiting 'general' esterases with DFP, an affinity gel
prepared by coupling a thiol analog containing the
trifluoropropanone derivative to epoxy activated
Sepharose allowed the specific binding of hemolymph JHE
activity to a small amount of gel. The gel could be
washed extensively to remove contaminating proteins.
Because of the extensive information available on
inhibitors of JHE, 3-octylthio-1,1,1-trifluoropropan-2-
one (OTFP) was chosen as a selective eluting agent. The
only serious difficulties in the purification were the
limited amounts of biological material available and the
long dialysis time needed to remove the eluting agent
(OTFP).[15,16]

This affinity purification permits one to obtain
useful amounts of pure enzyme from a very limited tissue
source. This enzyme, in turn, permits more

sophisticated biochemical characterization and an investigation of its biological activity. However, the most exciting aspects of characterization such as X-ray crystallography still require more protein than is readily available from insect hemolymph.

5. BIOCHEMICAL CHARACTERIZATION OF JHE

The biochemistry of the JHE's from several lepidopterous species have been well described given the limited protein that is available and have many features in common. They appear to be acidic proteins of about 60,000 molecular weight and are very stable to a variety of conditions that denature most proteins. Although their k_{cat} is not exceptionally high, the k_{cat}/K_m ratio is as high as would be expected of a scavenger enzyme whose concentration far exceeds the concentration of its substrate.[16,17]

Although the enzymes from different species have many similarities, there are also perplexing differences. The JHE of Manduca sexta is profoundly activated by organic solvents while that of Trichoplusia ni is inhibited and that of Heliothis virescens is capable of carrying out transesterification in the presence of some alcohols.[18] There are small differences in molecular weight and pI of isozymes present in the hemolymph and there are varying levels of glycosylation. At one extreme, at least 3 JHE isozymes appear glycosylated in T. ni, while one of two hemolymph isozymes appears glycosylated in H. zea, and there seems to be no glycosylation of JHE in H. virescens or M. sexta. Whether these isozymes are of biological importance remains to be seen.[19]

6. BIOLOGICAL ACTIVITY OF JHE

The observation that M. sexta deposits melanin in its epidermis when JH levels are lowered and thus turns from blue green to black has proven very useful in monitoring JHE.[20] Picomoles of affinity purified JHE from the hemolymph of M. sexta produces melanization at the next molt when injected into second, third, or fourth stadium larvae. The effect can be reversed by the topical application of juvenoids. A prepupal burst of JH appears necessary for successful pupation in the noctuiid caterpillar T. ni. A second bioassay for anti-JH agents has been devised based on this observation[21]

and injected JHE prevents pupation in these insects in a
dose dependent fashion.

These data are exciting from the standpoint of
providing additional evidence for the hypothesis that
the in vivo JH titer is reduced by a combination of
decreased biosynthesis and increased catabolism. They
also are interesting in showing that JHE is the first
biochemical anti-JH. However, attempting to control
insects with a material which is not active when
administered orally is foolish. Also despite the
attraction of using JHE as a biochemical anti-JH agent,
experiments with it will be limited if one must spend
hours bleeding precisely staged caterpillars to have
enough enzyme for each experiment.

7. CLONING OF JHE

Once a pure protein is acquired, it is obvious to
develop antibodies and determine the N-terminal sequence
of the material. Good antibodies were developed to the
JHE's of M. sexta, T. ni and H. virescens and the N-
terminal amino acid sequences determined. For a variety
of reasons H. virescens was targeted for an effort at
cloning JHE. These reasons include the simplicity and
lack of ambiguity of its N-terminal sequence, the high
specific activity and lack of glycosylation of the
enzyme, and of course the importance of this species as
a pest. A size selected expression library made from
fatbodies of L_5D_2 larvae of H. virescens was screened
first with an antibody and positives then were screened
with an oligonucleotide made to the N-terminal amino
acid sequence of the purified JHE. This procedure
resulted in 3 apparently full length cDNA clones of
approximately 3000 bp which had highly similar but not
identical sequences. Analysis of the predicted amino
acids of the affinity purified enzyme were determined by
Edman degradation.[22]

A complete nucleotide sequence of a cDNA of course
allows one to predict the complete amino acid sequence
of the protein. It was reassuring to see that the
cloned JHE's showed sequence homology to the insect and
vertebrate esterases so far cloned. All of the clones
had the serine motif of GLY-X-SER-X-GLY seen in all
serine esterases and proteases.[23] However, an exciting
observation by Blow and coworkers for serine proteases[24]
and recently confirmed by Smith et al.[25] did not appear
to be present (see below).

(ASP 102)COO$^{\ominus}$ —— HN(HIS57)N —— HO(SER 195)

Figure 4   The catalytic triad shown above
for serine proteases appears to be significantly altered in esterases. In JHE,
and possibly other esterases, a glutamate
may replace the aspartate. More importantly,
esterases show no conserved histidine in the
region of HIS 57 of α-chymotrypsin. Whether
this is replaced by another amino acid, such
as a highly conserved ARG-PHE motif, or by
different folding of the protein is not known.

The observation, from X-ray analysis of serine
proteases, that there was a histidine in the correct
position to hydrogen bond to the hydroxyl proton of
serine and transfer a proton to an adjacent aspartate,
explained how a relatively non-reactive serine could
attack the carbonyl of a peptide bond. Since esterases
including AChE had the same serine motif as proteases,
it was widely assumed that the same mechanism applied.
Sequence analysis of a variety of esterases indicates
that in JHE the aspartate motif is replaced by a
glutamate and that this replacement may occur in other
esterases. It also was noted that there was no
histidine corresponding to HIS 57 of chymotrypsin.
However, in esterases there was a highly conserved motif
surrounding an arginine at about the same location. It
is possible that esterases differ from proteases in
having a very different folding pattern allowing
histidine, conserved in remote stretches of the
molecule, to have catalytic involvement. However, it
also is possible that the catalytic triad is very
different from proteases possibly explaining why most
serine proteases have high esterase activity but why
esterases have weak to moderate activity on amide bonds.
It is difficult to envisage ARG with its very high pK_a
as a direct replacement for HIS in a proton transfer
reaction. However, the charged ARG may serve to orient
other amino acids or water in an analogous reaction.
This observation on esterases may indicate that the
protease model has not been appropriate for the design
for inhibitors of esterases such as AChE.[22]

8. BACULOVIRUS EXPRESSION OF JHE

A variety of arguments indicate that the cDNA's isolated in our laboratory actually code for JHE.[22] However, it would be reassuring to express those cDNA's *in vitro* and show that they have catalytic activity. It was noted many years ago that insect DNA viruses in the family Baculoviridae produced several high abundance proteins which did not appear critical for survival of the virus in the laboratory. Based on these observations, Summers and Smith[26] developed some highly efficient expression systems based upon the placement of a foreign gene coding sequences under the control of the polyhedron promoter. This expression system has proven very useful in many fields, and a variety of foreign proteins have been produced in insect cells at up to 500 mg per liter of culture medium.[27-30] A report by Maeda and coworkers[31] was exciting in that they demonstrated with a Bombyx mori system that the polyhedron promoter could be exploited for the production of foreign proteins *in vivo*. From the standpoint of confirming that the cDNA's isolated as described in section 7 actually coded for JHE and for the production of enough JHE for more sophisticated biochemical study, the baculovirus approach was very attractive.

A second attraction of the baculovirus system was the possibility of the development of a genetically engineered pesticide.[32] The most widely used virus for expression systems is the nuclear polyhedrosis virus (NPV) of Autographa californica, a noctuiid moth. This virus has a rather broad host range among many relatives of this pest including Heliothis, Trichoplusia, and Spodoptera sp. This and related viruses have been marketed as biological insecticides. However, they have failed to have a major impact in most developed countries for a variety of reasons.[33] Certainly one reason is that the insects usually are slow to die following viral infection.[34] During this time the insect continues to feed. This situation is advantageous when the virus is used as a biological control agent since it results in production of large amounts of virus. However, continued feeding is not advantageous when the virus is to be used as an insecticide.

Theoretical Advantages of Expression of JHE in NPV

The most obvious attribute of JHE is that it will result in anti-JH activity which is commercially desirable. Under laboratory conditions in certain species it may result in precocious metamorphosis, but a more likely result is a partial metamorphosis resulting in teratogenic effects. Since among many other effects, JH is a feeding stimulant, a still more likely result of JHE expression will be a reduction in feeding and rate of growth of the larva. Obviously each of these effects is attractive for insect control in reducing crop damage. It also is possible that JHE expression will reduce feeding damage from insects receiving a sublethal viral infection. This could result in a wider effective host range or lower levels of viral application in the field.

At least in part, commercial production of viruses will probably be in insects, for at least the near future. If the virus causes insects to die early or to grow slowly, production will be difficult. In theory the effects of precocious JHE can be reversed by the application of juvenoids which lack an ester moiety easily hydrolyzed by JHE. Thus viruses can be produced in normal sized or even giant insects. Such commercial uses of juvenoids already exist in the silk industry.

The enzyme also offers the advantage over lytic toxins in that it is toxic to the organism as a whole, but not to the tissues producing it. As a natural enzyme of the target insects, it was also hoped that the biochemical machinery to produce it in large amounts would be present. Along this line the 19 amino acid leader sequence of JHE should result in its rapid export into the hemolymph. Thus the enzyme would appear in the very *in vivo* pools where it normally occurs in the last larval stadium.

Since JHE is a natural enzyme which normally occurs in the pest insects which it would be used to control, it may be less offensive to environmental groups opposing the release of some genetically engineered organisms. Along this line, the highly sensitive and specific assays for JHE in conjunction with antibodies and cDNA probes will make it easy to monitor in the field. Thus, even if it lacks commercial biological activity, JHE may represent a useful marker in model viruses for ecological studies. Although this would have to be tested, JHE is thought to lack biological activity in organisms not using JH and even to be inactive when orally administered to caterpillars.

Although baculoviruses and related insect viruses
represent the most obvious way to exploit the JHE gene,
the esterase from H. virescens should be active in any
species which uses a JH with a 2,3-E methyl ester to
control development or reproduction. Thus, the gene
should be attractive for expression in a variety of
systems.

Construction of Transfer Vector

The transfer vectors available from the Institute
of Virology had either Bam HI or Bgl II cloning sites,[35]
while the three JHE clones all were in Eco RI sites in a
Bluescript plasmid. Since the JHE gene had an internal
Bam HI site, a Bgl II linker was placed at the 5' end of
the gene. The plasmid was cut at a Cla I site just
downstream from the natural stop codon, the ends
repaired with the Klenow fragment of the DNA polymerase
from E. coli and a second Bgl II linker inserted. The
entire coding region lacking the 3' noncoding region
thus could be removed and inserted into Bam or Bgl sites
in the existing vectors. The JHE gene was first
inserted in vector RP 23 in which the foreign gene is
placed under the control of the polyhedron promoter and
no polyhedron protein is made.

Expression of JHE

When cells of Spodoptera frugiperda were co-
transfected with this transfer vector and the wild type
virus, JHE activity could be detected even in the crude
transfection mixture. Plaque assays were carried out by
screening for polyhedron minus plaques. Each of the
polyhedron minus plaques picked had high JHE activity.
After 5 sequential plaque purifications, the polyhedron
minus-JHE plus virus was considered to be genetically
homogeneous. In spinner flask culture, cells of S.
frugiperda have JHE activity below detectable limits.
When the virus is added with a multiplicity of infection
of 10 plaque forming units per cell, JHE activity in the
medium increases rapidly from about 10 hours to a
maximum at 48 hours followed by a slow decline. As
expected with the amino acid leader sequence, the JHE
activity appears predominantly in the supernatant rather
than the cells. The catalytic activity in the medium
was about 100 nmol/min-ml which is about twice the
activity normally seen in the hemolymph of H. virescens
at optimum production of JHE and corresponds to about 75
mg/liter of media.

Biological Activity of Expressed JHE and the Engineered Virus

When the expressed JHE from a spinner culture was partially purified by open column and HPLC anion exchange chromatography and injected into second stadium larvae of M. sexta, the same dose dependent blackening was seen. At lower doses, the insects permanently stopped feeding without a color change.

Feeding of the polyhedron minus virus containing the JHE gene to first stadium larvae of T. ni resulted in viral infections and very high mortality about 8 days post infection. Within 3 days of infection many larvae were observed to feed at a greatly reduced rate. Many of these larvae remained in the third stadium while the untreated control larvae reached pupation. When the larvae fed at a variety of doses of the virus were weighed 6 days post infection, control larvae weighed an average of 178 ($\pm$ 68) mg and larvae exposed to diet containing 3.107 plaque forming units/sq inch of medium surface weighed 37 ($\pm$ 14) mg each. There was a clear dose response between viral exposure and reduced weight. Some of these small larvae lived for over two weeks following infection. Since some of these effects can be reversed by the application of a juvenoid, it appears that the expressed JHE is responsible for at least some of the reduction in growth. The biology of the insect-virus interaction appears complex in terms of a number of diverse biological effects being observed. A variety of studies will be needed to determine which effects are due to local vs. systemic effects of the expressed JHE and which effects are due to JHE and which to the virus.

By 4 days post infection, the JHE levels in the hemolymph of infected larvae were over 4 times that of control larvae. Six days after infection the hemolymph from apparently stunted animals was examined and found to have the level seen in most control feeding larvae and about the same activity as a burst of JHE activity seen at ecdysis. However it is only about 10% the activity seen in the major JHE peak in the last larval stadium of T. ni and 10% of the activity seen in the cell culture media.

Limitations of Expressed JHE for Insect Control

From the standpoint of ultimately using recombinant DNA technology for the design of pathogens effective in insect control, the above data are very encouraging.

However, there are many potential limitations. The most obvious is that reducing feeding damage while the virus kills the insect is advantageous, but it approaches only one of many problems that have restricted the use of baculoviruses in the field. Possibly this success will encourage additional investment in problems such as formulation and production.

In addressing the specific problem of what gene to insert into viruses to speed kill, several criteria should be considered. Certainly JHE has a theoretical advantage over lytic toxins which may kill infected cells before sufficient toxin is released for a systemic effect. However, a major disadvantage of enzymes like JHE is that they are very large compared to many small toxins and neuropeptides. It will be interesting to see whether a small molecule acting stoichiometrically or a large molecule acting catalytically is more efficient.

It also should be realized that the insect probably has many ways to overcome the precocious expression of JHE. First, metabolism of JH can occur by several pathways, and it does occur even in intermolt periods in early stadia. Thus, the expressed JHE certainly increases the rate of JH degradation, but it is increasing at a rate over existing degradation. Animals also normally have numerous control mechanisms. The caterpillar could increase the biosynthesis of JH or alter receptor density to overcome some increased catabolism. We have found that about 5% of the JH present at the start of the last larval stadium is sufficient to delay wandering and continue feeding. Thus, the expressed JH must be very efficient at removing the last traces of JH and a slight increase in JH biosynthesis or even remethylation may be able to overcome the catabolism caused by JHE. For these reasons it would be desireable to have in vivo levels of expressed JHE activity much higher, rather than much lower, than natural levels seen in the last larval stadium.

Possibly the most disturbing observation regarding JHE was its very short half-life in vivo in spite of exceptional stability to a variety of very harsh conditions in vitro. Thus, it seems certain that the insect is capable of removing JHE from its hemolymph. When one examines the physiology of noctuiid caterpillars, it becomes clear that the esterase's half-life must be very short. Possibly the insect can harness these natural mechanisms to overcome the precocious expression of the enzyme.

A clear goal of this work is to reduce feeding damage, however, there will be a series of delays. There will be a delay from the time that the JH titer is reduced until feeding stops and certainly a delay between the release of JHE in the hemolymph and effective reduction of JH in the tissues. With only moderate levels of JHE expression, there will be a delay before sufficient cells are infected to produce biologically effective levels of the enzyme. Finally, the polyhedron promoter is activated very late in viral infection, further delaying the appearance of JHE. With all of these sequential delays in the system, will the biological effects occur soon enough to be of commercial interest?

Finally, the data reported here are from a severely attenuated polyhedron minus virus. Although the insertion of the JHE gene clearly improves the effectiveness of this virus in reducing insect feeding under the bioassay conditions used here, it may not improve the effectiveness of wild type viruses. Certainly, any foreign gene inserted into a natural virus has the possibility of reducing the effectiveness of that virus. It will be interesting to see if the added value of JHE will compensate for the limitations caused by the expression of a foreign protein.

Future Research

Although far from a panacea, the dramatic reduction in feeding caused by JHE in the polyhedron minus baculovirus is very encouraging. Certainly many of the limitations discussed above can be approached. The fact that biological activity is seen at levels of expression below that observed naturally in vivo or that reported for maximum expression of foreign proteins in vitro is encouraging. There certainly is room for improvement of expression by a variety of mechanisms.

A detailed study of the stability of both the protein and the corresponding mRNA following in vivo and in vitro expression may provide leads on how to stabilize these two biopolymers to increase production of enzyme activity in vivo. The fact that the insect is known to clear both mRNA and protein very quickly following maximal levels of esterase activity, indicates that there may be rapid turnover mechanisms. Information on what controls mRNA lifespan is increasing rapidly.[36] At the protein level removal of sites labile to hydrolysis by site directed mutagenesis is reasonable. An interesting possibility is that if insects have a specific mechanism for the removal of JHE, we may find that a JHE gene isolated from a very distantly related species will be more active in controlling noctuiid caterpillars than one isolated from H. virescens or closely related species.

Certainly continued work on promoter systems which retain essential viral proteins, which produce large amounts of the foreign protein, and which initiate production early in the viral cycle is essential. The fact that JHE activity can be monitored quantitatively and specifically on very small samples indicates that this gene may be a very useful reporter gene for such studies.

As mentioned earlier, the JH titer of caterpillars is reduced by a decline in biosynthesis of JH and an increase in degradation by JHE. The virus reported here only results in an increase in degradation. An obvious approach as multipromoter systems become available, would be to include a gene incoding for an allatostatin as well as JHE to reduce biosynthesis of JH as well as to increase degradation. The virus would then be mimicking closely the JH regulatory mechanisms in the natural insect. A more distant project would be to target the factors which regulate both JHE and allatostatin to initiate the pupation sequence.

In some insects very low levels of JH are required to maintain active feeding. If there are adequate feed back mechanisms to the corpora allata, JHE might be unable to overcome biosynthesis in some species. As multipromoter systems become available, it may be possible to overcome this difficulty by expressing an allatostatic factor as well as JHE. Such a system would mimic the insect endocrine system even more closely by a combined reduction of biosynthesis and increased degradation.

Overall the results are very encouraging for molecular approaches to insect control. Far better agents to express in baculoviruses and other vectors may become available soon, yet the demonstration that JHE is effective in this limited test does demonstrate that the approach is feasible. However, these data should pose a warning to environmental chemists and toxicologists who have spent a generation perfecting means to detect and determine risk of small molecules. It is time that we began to train individuals to handle the toxicology and environmental chemistry of recombinant molecules. They are exciting as they expand the repertoire of control agents available to agricultural and medical entomology. However, these recombinant molecules will be used with an insecticide philosophy. Without adequate data to assure the public on the safety of these materials, they may face many of the same real and perceived problems as existing compounds.

9. INTEGRATION OF MOLECULAR AND CHEMICAL APPROACHES

The above academic exercise demonstrates a useful interaction between chemical and molecular approaches to insect control. The production of a recombinant baculovirus that reduces feeding damage is exciting, but this work could not have been done without the chemical background demonstrating the in vivo role of the enzyme and providing the affinity chromatography systems for the isolation of sufficient protein to prepare antibodies and determine a partial sequence. The nucleotide sequence, in turn, has provided insight into the catalytic mechanism of esterase action which may be useful for the future design of new inhibitors.

The large number of compounds that are active in screening tests but do not lead to structures that are of commercial activity, represent a tremendous waste in the agricultural chemical industry. A severe limitation of SAR type approaches is that they allow rapid optimization of structures in a compound series, but do not allow one to jump to radically different structures that may attack the same target.

Currently when one is unable to obtain commercial activity in a series of compounds by SAR and rational design based on a lead from screening, the compounds and the tremendous investment in the series are lost. However, at the very least, the screening effort has defined a new biochemical target which if disrupted would result in an agriculturally desirable phenomenon. This observation provides the perfect lead to molecular biologists who at this point know of very few genes which if over or under produced would lead to a commercial product.

In the past it has been hard to justify biochemistry on a target since that biochemistry may not

lead directly to a product. With the advent of
molecular biology, biochemical studies on the mechanism
of action of a compound can lead to exploitation by
approaches both in chemistry and recombinant DNA
technology. The fact that biological and biochemical
approaches can lead directly to a product should make
them financially much more attractive.[4]

The process of molecular exploitation of biological
activity relies on the development of immunochemical or
nucleic acid probes to isolate the target gene which in
turn relies on the isolation of the protein. In the
past such protein isolation could take many man years.
However, in the hypothetical case outlined here, one
already knows from SAR approaches, a variety of
molecules which interact with the target protein. Since
a series of molecules of varying activity is known, a
variety of affinity columns of increasing affinity can
be prepared and a selection of elution compounds already
are available.

The pure protein not only would provide the in
vitro kinetic or binding assays that have been used in
structure optimization in the past, but provide a route
for molecular approaches which may be able to lead
directly to a product based on the initial lead from
screening of synthetic chemicals. Of course the base
sequence of the message or gene which would come out of
the recombinant work would provide an amino acid
sequence that could be of use in compound design, and
expression of large amounts of protein could lead to a
further definition of the site by physical techniques
including X-ray crystallography.

Clearly, an X-ray structure does not lead directly
to new biologically active compounds, but with computer
aided design programs it does provide a tool to direct
creative synthetic efforts. The fact that the
biochemistry is facilitated by previous investments in
synthesis and that the biochemistry can be justified by
direct exploitation by molecular biology as well as
creative synthesis, makes the approach much more
attractive.

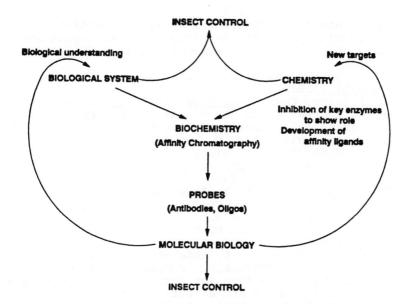

Figure 5   For enhanced rates of discovery of biological activity it must be recognized that the fields of chemistry and biotechnology can support one another as well as drive directly toward commercial products in their own right.

Certainly it seems reasonable that modern techniques in synthetic chemistry as well as molecular biology can be seen as complementary rather as competing technology.   Hopefully such integration will increase the rate of success and decrease the cost of discovery of insecticides, herbicides, and other chemicals of biological importance.[11]

REFERENCES

1.   R. L. Metcalf, <u>Annu. Rev. Entomol.</u>, 1980, <u>25</u>, 219.
2.   B. D. Hammock and D. M. Soderlund, 'Pesticide Resistance, Strategies and Tactics for Management', (E. Glass, ed.) National Academy of Sciences Press, Washington D.C., 1986, p. 111.
3.   C. A. D. de Kort and N. A. Granger, <u>Annu. Rev. Entomol.</u>, 1981, <u>26</u>, 1.

4.   B. D. Hammock, 'Comprehensive Insect Physiology
     Biochemistry and Pharmacology', (G. A. Kerkut and
     L. I. Gilbert, eds.) Pergamon Press, N. Y., 1985,
     p. 431.
5.   T. C. Sparks and B. D. Hammock, Pestic. Biochem.
     Physiol., 1980, 14, 290.
6.   L. Pauling, Chem. Eng. News, 1946, 24, 1375.
7.   P. R. Andrews and D. A. Winkler, 'Drug Design: Fact
     or Fantasy', (G. Jolles and K. R. H. Wooldridge,
     eds.) Academic Press, N. Y., 1984, p. 145.
8.   B. D. Hammock, K. D. Wing, J. McLaughlin, V. M.
     Lovell, and T. C. Sparks, Pestic. Biochem.
     Physiol., 1982, 17, 76.
9.   B. D. Hammock, Y. A. I. Abdel-Aal, C. A. Mullin, T.
     N. Hanzlik, and R. M. Roe, Pestic. Biochem.
     Physiol., 1984, 22, 209.
10. M. M. Olmstead, W. K. Musker, and B. D. Hammock,
     Acta Cryst., 1987, C43, 1726.
11. B. D. Hammock, Y. A. I. Abdel-Aal, M. Ashour, A.
     Buehler, T. N. Hanzlik, R. Newitt, and T. C.
     Sparks, 'Human Welfare and Pest Control Chemicals -
     Approaches to Safe and Effective Control of Medical
     and Agricultural Pests', (M. Sasa, S. Matsunaka, I.
     Yamamoto, and K. Ohsawa, eds.), Nissan Science
     Foundation, Tokyo, 1986, p. 53.
12. Y. A. I. Abdel-Aal and B. D. Hammock,
     'Bioregulators for Pest Control', (P. A. Hedin,
     ed.), ACS Symposium Series 276, Washington D. C.,
     1985, p. 135.
13. A. Székács, B. D. Hammock, Y. A. I. Abdel-Aal, P.
     P. Halarnkar, M. Philpott, and G. Matolcsy, Pestic.
     Biochem. Physiol., 1989, 33, 112.
14. A. Székács, B. D. Hammock, Y. A. I. Abdel-Aal, M.
     Philpott, and G. Matolcsy, 'Biotechnology for Crop
     Protection', (P. A. Hedin, J. J. Menn, and R. M.
     Hollingworth, eds.), ACS Symposium Series 379,
     Washington D. C., 1988, p. 215.
15. Y. A. I. Abdel-Aal and B. D. Hammock, Science,
     1986, 233, 1073.
16. T. N. Hanzlik and B. D. Hammock, J. Biol. Chem.,
     1987, 262, 13584.
17. Y. A. I. Abdel-Aal, T. N. Hanzlik, L. G. Harshman,
     B. D. Hammock, and G. Prestwich, Comp. Biochem.
     Physiol., 1988, 90B, 117.
18. G. Croston, Y. A. I. Abdel-Aal, S. J. Gee, and B.
     D. Hammock, Insect Biochem., 1987, 17, 1017.
19. B. D. Hammock, Y. A. I. Abdel-Aal, T. N. Hanzlik,
     G. E. Croston, and R. M. Roe, 'Molecular
     Entomology', (J. Law, ed.) A. R. Liss, N. Y., 1987,
     49, 315.

20. J. W. Truman, L. M. Riddiford, and L. Safranek, <u>J.</u> <u>Insect Physiol.</u>, 1973, <u>19</u>, 597.

21. T. C. Sparks, R. M. Roe, A. Buehler, and B. D. Hammock, <u>Insect Biochem.</u>, 1987, <u>17</u>, 1011.

22. T. N. Hanzlik and B. D. Hammock, <u>J. Biol. Chem.</u>, 1989, in press.

23. S. Brenner, <u>Nature</u>, 1988, <u>334</u>, 528.

24. D. M. Blow, J. J. Birktoft, and B. S. Hartley, <u>Nature</u>, 1969, <u>221</u>, 337.

25. S. O. Smith, S. Farr-Jones, R. G. Griffin, W. W. Bachovchin, <u>Science</u>, 1989, <u>244</u>, 961.

26. M. D. Summers and G. E. Smith, <u>Texas Agricultural Experiment 29.Station, Bull.</u>, 1987, <u>1555</u>, 1.

27. L. K. Miller, <u>Annu. Rev. Microbiol.</u>, 1988, <u>42</u>, 177.

28. V. A. Luckow and M. D. Summers, <u>Biotechnology</u>, 1988, <u>6</u>, 47.

29. D. H. L. Bishop and R. D. Possee, <u>Adv. in Gene Technology</u>, 1989, in press.

30. S. Maeda, <u>Annu. Rev. Entomol.</u>, 1989, <u>34</u>, 351.

31. S. Maeda, T. Kawai, M. Obinata, H. Fujiwara, T. Horiuchi, Y. Saeki, Y. Sato, and M. Furusawa, <u>Nature</u>, 1985, <u>315</u>, 592.

32. J. B. Kirshbaum, <u>Annu. Rev. Entomol.</u>, 1985, <u>30</u>, 51.

33. J. Huber, 'The Biology of Baculoviruses', (R. R. Granados and B. A. Federici, eds.) CRC Press, Boca Raton FL, 1986, <u>II</u>, p. 181.

34. G. A. Benz, 'The Biology of Baculoviruses', (R. R. Granados and B. A. Federici, eds.) CRC Press, Boca Raton FL, 1986, <u>I</u>, p. 1.

35. Y. Matsuura, R. D. Possee, H. A. Overton, and D. H. L. Bishop, <u>G. Gen. Virol.</u>, 1987, <u>68</u>, 1233.

36. T. Hunt, <u>Nature</u>, 1988, <u>334</u>, 567.

Chemical Effectors of the Ryanodine Receptor: A Novel Strategy for Insect Control

Isaac N. Pessah

DEPARTMENT OF VETERINARY PHARMACOLOGY AND TOXICOLOGY, UNIVERSITY OF CALIFORNIA, DAVIS, CALIFORNIA 95616, USA

I. Introduction

Since it has become apparent in recent years that the radioligand [^{3}H]ryanodine serves as a highly selective conformational probe for various functional states of the calcium (Ca^{2+}) release channel protein of muscle sarcoplasmic reticulum (SR), I will present data which explores mechanisms by which chemically diverse groups of substances influence the binding of [^{3}H]ryanodine to the Ca^{2+} release channel and thereby alter its function. This paper, therefore, is not intended to be a review of what is currently known about the structure of the ryanodine receptor (recent articles exploring ryanodine receptor structure may be found in references 50 and 51). The significance of the ryanodine receptor as a potential target for insect control is underscored by its key role in the process of excitation-contraction (EC) coupling in striated muscle. Therefore, [^{3}H]ryanodine may prove invaluable for identifying classes of compounds which directly influence EC coupling at the level of the SR Ca^{2+} release channel. EC coupling has not been previously exploited as a biochemical target for insecticide development and what makes this strategy most attractive is the fundamental essentiality of the process for the normal operation of muscle. Ultimately, rational design of agents which selectively alter the integrity of EC coupling can influence every life stage of a pest species. The emphasis of this paper is on the allosteric nature of

Supported by NIH Grant ES05002

the [³H]ryanodine receptor complex and the mechanisms by which chemical modifiers influence receptor function. Although much of the data presented is obtained from vertebrate muscle, preliminary data is presented from crayfish muscle.

II. An Overview of EC Coupling

Insect muscles share many structural features with vertebrate striated muscle. Although all insect muscles exhibit striation, there exists a great deal of diversity in their ultrastructural detail and

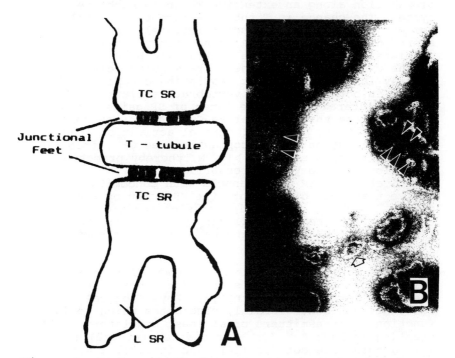

Figure 1 A; Longitudinal representation of a triad junction showing the relative position of the T-tubule membrane, the terminal cisternae (TC) of SR, the longitudinal (L) SR and the junctional feet processes. B; TC SR membrane vesicles stained with uranyl acetate showing junctional feet processes (dark arrows); x121,000. Micrograph courtesy of Dr. D. J. Scales.

physiology. Insect muscles have been extensively
studied (1, for review), but as is the case with
vertebrate muscle, the least understood steps in the
contraction-relaxation cycle are the molecular
mechanisms taking place during EC coupling. EC
coupling in insect and vertebrate striated muscle
cells occurs at the triad junction (or more typically
the diad junction in insect muscle) where the
transverse (T) tubule membrane is positioned only 100-
150 Å from the terminal cisternae (TC) of SR (Figure
1A). The action potentials at the surface membrane
(i.e., the sarcolemma) of the typical striated muscle
cell are brought deep into the muscle fiber by
propagation along an extensive T tubule membrane
system. The change in T tubule membrane potential
somehow signals the release of intracellular Ca^{2+}
stored within the TC of SR; a necessary requisite for
complete activation of the contractile proteins found
in the cytoplasm. Hence the intracellular release
pathway is a key component regulating cytoplasmic Ca^{2+}
levels in both insect and vertebrate striated muscle.
Ca^{2+} sequestered within the lumen of SR gains access to
the cytoplasm through putative Ca^{2+} release channels
localized at the TC of SR. Channel activation leads to
a rapid elevation the sarcoplasmic free Ca^{2+}, from $<10^{-7}$
M in the resting state to $>10^{-6}$ M, thereby promoting
formation of actomyosin crossbridges (2-4).
Inactivation of SR Ca^{2+} channels with concomitant
active transport of Ca^{2+} into the SR lumen by the
$Ca^{2+}(Mg^{2+})$-ATPase, in addition to electrogenic Na^+/Ca^{2+}
exchange with the extracellular space in cardiac
fibers, restores the resting Ca^{2+} level in the myoplasm
(5-6).

Considerable attention is being focused on the Ca^{2+}
release channels of SR in search for a possible
chemical link in EC coupling (7). Physiologically-
relevant, yet controversial, candidates which could
function as transmitters or at least modulators in EC
coupling include Ca^{2+} (Ca^{2+}-induced Ca^{2+} release) (8-
13), adenine nucleotides (14), and inositol 1,4,5-
trisphosphate (IP_3) (15-16). In addition, Ca^{2+}
permeability of SR in intact or skinned muscle fibers
as well as purified SR vesicles enriched in membranes
of TC origin can be altered with a large number of
chemically heterogeneous compounds including ryanoids
(17-27), methyl xanthines (28-33), anthraquinones (33,
34), local anesthetics (27, 35-36), miconazol (27),
heavy metals (37-39), aryldisulfides (27, 40),

dantrolene (41-43), antibiotics (44), and ruthenium red (11, 45). Many of these compounds exhibit antithetical effects on SR Ca^{2+} permeability depending not only on experimental conditions but also on the type of muscle studied thereby confounding mechanistic conclusions. To understand whether these diverse compounds exert inotropic effects on muscle by influencing unrelated and biochemically distinct sites or by acting on a single channel gating mechanism, elucidation of those mechanisms, and isolation and structural characterization of their binding sites requires a direct means of assessing Ca^{2+} channel structure and function.

Implementation of rotary shadowing and negative stain techniques yield ultrastructural details of the triad junction and especially of the junctional face of SR.

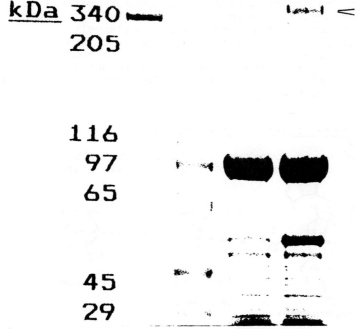

Figure 2 SDS-PAGE of purified SR membrane fractions of longitudinal origin (lane 3) and of junctional origin (lane 4). Junctional foot protein enriched in the junctional membranes is indicated by the arrow. Lanes 1 and 2 show molecular weight standards.

These studies convincingly demonstrate the presence of
discrete rows of structures termed junctional feet
each having an approximate diameter of 25 nm which
bridge the T tubule-SR membranes at the triad (46-48;
Figure 1B). The junctional feet structures have been
suspected for many years to be responsible for
conveying the information of the T tubule action
potential to the release of Ca^{2+} from the TC SR. Direct
evidence of an involvement has been provided only
within the last 4-5 years due to the availability of
[^{3}H]ryanodine of high specific activity for receptor
studies (49).

Recognition that four polypeptides, each having a
molecular mass >400 kDa (Figure 2), not only
constitutes the high molecular weight oligomer of the
junctional foot but also comprises the [^{3}H]ryanodine
receptor and the Ca^{2+} release channel, has permitted
significant advances in our understanding of the
structure and function of a key component critical to
excitation-contraction coupling in muscle (50). Recent
cloning of the ryanodine receptor conclusively
demonstrates its association with the junctional foot
protein (51). These findings not only open new areas
of fundamental research of insect muscle but also
allow rational design of insecticides having a novel
mechanism of action.

III. The Mode of Action Of Ryanoids

Waterhouse and coworkers identified the alkaloid
9,21-didehydroryanodine (DRY) as an active principle

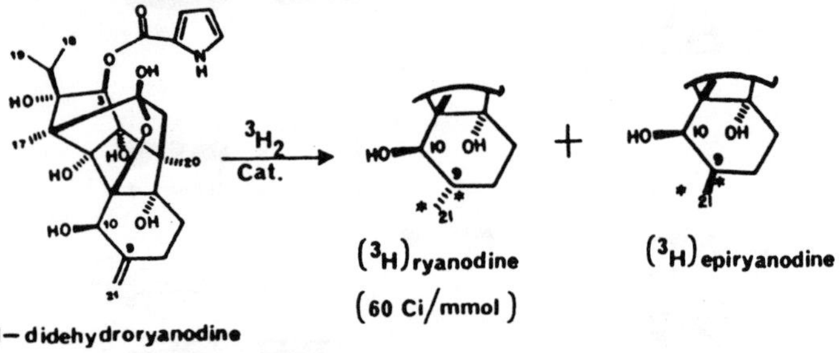

9,21-didehydroryanodine

^{3}H$_2$
Cat.

(^{3}H)ryanodine
(60 Ci/mmol)

(^{3}H)epiryanodine

Figure 3 Synthetic scheme showing the reduction of
9,21-didehydroryanodine with tritium gas to yield
[^{3}H]ryanodine of high specific activity, which is
easily separated from its epimer by HPLC.

in commercial preparations of ryanodine and in the
bark of the tropical shrub <u>Ryania speciosa</u> (Figure 3)
(52-53). In addition to having biological potency
equivalent to ryanodine, the unsaturation at the 9,21-
position permitted a single-step tritiation yielding
[9,21-^{3}H$_2$]ryanodine of high specific activity (60
Ci/mmol) for use in receptor studies (50; Figure 3).

[^{3}H]Ryanodine binding sites are unmasked in the
presence of micromolar Ca^{2+} (<u>i.e.</u>, binding is Ca^{2+}
dependent) in both vertebrate (49) and invertebrate
preparations (Figure 4), and is enhanced by chemical
agents which promote release of Ca^{2+} from SR, including

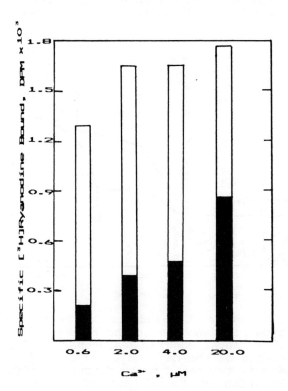

<u>Figure 4</u>   Binding of 5nM [^{3}H]ryanodine to heavy SR
fraction from crayfish abdominal muscle as a
function of Ca^{2+} in the absence (shaded bar) and
presence (open bar) of 20 mM caffeine.

adenine nucleotides, caffeine, and anthraquinones
(discussed below, Sections V and VI) (40, 50, 54-55).
Ligands which inhibit Ca^{2+} release from SR such as Mg^{2+}
and ruthenium red, also inhibit the activation of
[^{3}H]ryanodine binding sites by Ca2+ (discussed below,
Section VII).

Figure 5 shows typical Ca^{2+} transport measurements
from vertebrate SR vesicles which further demonstrate
the Ca^{2+}-dependent nature of ryanodine action on the
release channel. Vesicles are actively loaded with Ca^{2+}
in the presence of ATP, and release is induced by a
bolus (25 µM) of Ca^{2+} (*i.e.*, Ca^{2+}-induced Ca^{2+} release).
These results in conjunction with [^{3}H]ryanodine
receptor binding studies strongly suggest that
nanomolar ryanodine binds to the Ca^{2+}-activated

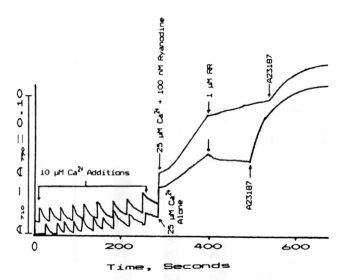

Figure 5 Ca^{2+} uptake, Ca^{2+}-induced Ca^{2+} release, and
the influence of ryanodine on the rate of release
from rabbit skeletal muscle. Measurements were made
in the presence of the Ca^{2+} indicator antipyrylazo
III. Release is triggered by an abrupt increase in
Ca^{2+} (25 µM) which in the presence of 100 nM
ryanodine yields significantly enhanced rates of
release. Ca^{2+} release under both conditions is
inhibited by 1 µM ruthenium red. Residual vesicular
Ca^{2+} is released by the ionophore A23187.

open state of the Ca^{2+} release channel thereby
preventing its complete inactivation. Since
dissociation of ryanodine from its receptor site is
extremely slow (40), an immediate consequence of
receptor binding is prolonged activation of the
contractile elements in the sarcoplasm and failure of
the affected muscle fibers to relax. The mechanistic
scheme summarized in Figure 6 may not only account for
ryanodine's toxicity but also suggests mechanisms by
which allosteric modulators of the ryanodine receptor
may influence channel function.

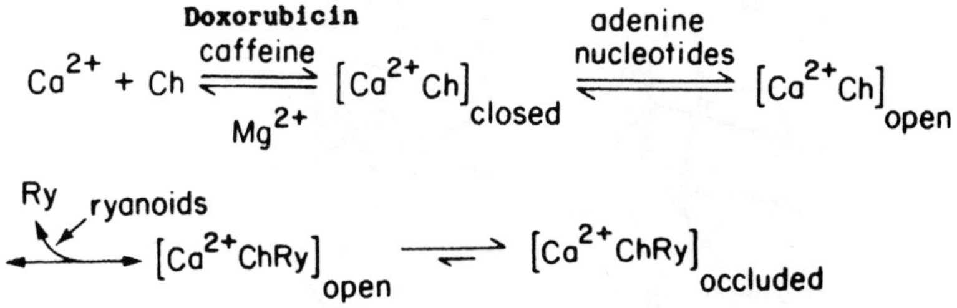

$$Ca^{2+} + Ch \underset{Mg^{2+}}{\overset{\substack{\textbf{Doxorubicin} \\ \text{caffeine}}}{\rightleftharpoons}} \left[Ca^{2+}Ch\right]_{closed} \overset{\substack{\text{adenine} \\ \text{nucleotides}}}{\rightleftharpoons} \left[Ca^{2+}Ch\right]_{open}$$

$$\overset{Ry \ \ \text{ryanoids}}{\longleftarrow} \left[Ca^{2+}ChRy\right]_{open} \rightleftharpoons \left[Ca^{2+}ChRy\right]_{occluded}$$

Figure 6   Mechanistic scheme of the interaction of
ryanodine with its receptor in muscle. The influence
of various chemical modulators on the reaction
sequence is shown.

IV. Ryanoid Structure-Activity Relationships

Unfortunately the complexity of the ryanoid molecule
has limited structural modifications to a handful of
alkylation and degradation reactions which invariably
yield products having much reduced toxicity and
receptor activity (56). Two notable exceptions are the
hydrolytic products ryanodol and 9,21-
didehydroryanodol (Figure 7). Despite their complete.
lack of toxicity to mice and their very low activity
in competition with [^{3}H]ryanodine at the vertebrate
receptor, these compounds maintain a high degree of
toxicity to house flies and American cockroach (e.g.,
piperonyl butoxide-synergized KD_{50} in house flies =
0.34 $\mu g/g$) (56). The mechanism for this remarkable
selectivity has not been determined, however one
cannot yet discount the possibility of major
differences in the structural requirements of the
invertebrate receptor.

V. Compounds Which Sensitize The Channel To Activation

Xanthines such as caffeine, theophylline, and enprofylline at high concentrations (typically above millimolar) allosterically influence the sensitivity of the typical vertebrate [^{3}H]ryanodine binding site to activation by Ca^{2+} (40). This influence of the xanthines is also apparent with the invertebrate

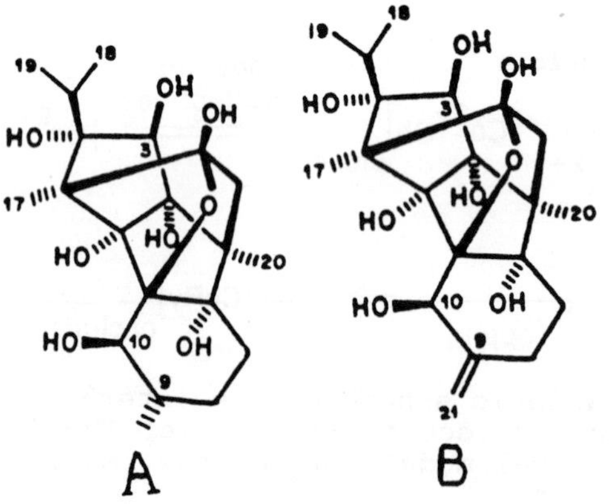

Figure 7 The structure of two ryanoids ryanodol (A) and 9,21-didehydroryanodol (B) which are highly insecticidal yet showing low mammalian toxicity.

receptor (Figure 4) where the presence of 20 mM caffeine increases the sensitivity of the ryanodine receptor to activation by Ca^{2+} by approximately 20-fold. Anthraquinones, such as doxorubicin, have been found to have similar properties on the ryanodine receptor with the exception of being over 3 orders of magnitude more potent than caffeine (57). The relative potency of two anthraquinones, doxorubicin and daunorubicin, and caffeine for activating the ryanodine receptor from vertebrate skeletal muscle at low Ca^{2+} are illustrated in Figure 8. These observations may serve as a basis for rational development of synthetically-accessible chemical structures having selectivity towards the insect ryanodine receptor.

VI. Compounds Which Influence Receptor Occupancy

Adenine nucleotides and their xenobiotic congeners enhance [^{3}H]ryanodine receptor occupancy and kinetic rates of association thereby influencing the intensity of the response from Ca^{2+} release channels to a given level of trigger Ca^{2+} (14, 40).The mechanism responsible appears to involve an adenine nucleotide-induced stabilization of the open channel conformation (14, 40; Figure 6). Significant synergism is observed with simultaneous exposure of the ryanodine receptor to xanthine and adenine nucleotide, and suggests a cooperative functional response of the Ca^{2+} channel to changing metabolic levels of ATP, xanthine, and pH. Xenobiotic analogs of adenine (as well as other) nucleotides may provide valuable information about the details of these interactions.

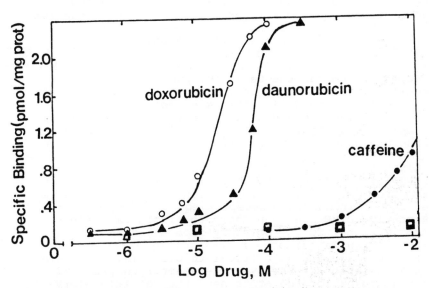

Figure 8   Anthracycline- or caffeine-induced sensitization of [^{3}H]ryanodine-binding to skeletal TC SR membranes (50 μg protein) in buffer containing 2 μM free Ca^{2+} (adjusted with EGTA), 1mM free Mg^{2+}, 2.5nM [^{3}H]ryanodine.

VII. Compounds Which Inhibit Receptor Occupancy

In addition to unlabelled analogs of [^{3}H]ryanodine, a
number of compounds inhibit radioligand binding.
Ruthenium red, a polycationic dye, inhibits Ca^{2+}
release (Figure 5) and is a potent (IC_{50} = 40-500 nM,
depending on assay conditions) competitive inhibitor
of the [^{3}H]ryanodine binding site (49). Analogs of
ruthenium red which inhibit the ryanodine receptor
have recently been reported (58).

Redox reactions between specific sulfhydryl moieties
of cysteine within or near the Ca^{2+} regulatory domain
of the ryanodine receptor are apparently influenced by
Ca^{2+}-induced changes in protein conformation and may
play a critical role in gating the Ca^{2+}-release
channel. Abramson and coworkers (59) have demonstrated
that micromolar levels of Hg^{2+}, Ag^+, or Cd^{2+} induce
rapid calcium release from skeletal SR vesicles.
Ag^+-induced Ca^{2+} release is associated predominantly
with heavy fractions enriched in elements of the TC,
is comparable to the rate of release and tension
development in muscle fibers, is blocked by ruthenium
red or high levels of Mg^{2+}, and appears to proceed by
sulfhydryl oxidation not involving the Ca^{2+}-ATP_{ase}
(60-63). [^{3}H]ryanodine binding to SR membranes is
inhibited by Cd^{2+} and Ag^+ at concentrations below μM
(40, 49). Whereas the Ca^{2+}- [^{3}H]ryanodine-receptor
equilibrium complex is only slowly reversed by excess
unlabelled ryanoid (40, 55), micromolar Ag^+ or
aryldisulfide rapidly dissociate the ligand-receptor
complex (40, Figure 9). Aryldisulfides react with
absolute specificity with protein thiols via a
disulfide interchange. 5,5'-Dithiobis(2-nitro)benzoate
(DTNB) as well as other aryldisulfide at
concentrations below 1 μM specifically derivatizes
receptor thiols with a concomitant rapid dissociation
of Ca^{2+}-[^{3}H]ryanodine- receptor equilibrium complex.
Recently, DTNB has been shown to elicit Ca^{2+} release
from skeletal TC SR, possibly from the same channel
involved in Ca^{2+}-induced release of Ca^{2+} (27). These
results suggest that critical thiols remain accessible
upon formation of the ryanoid-receptor complex and
appear to be located at or near the Ca^{2+} regulatory
domain of the channel protein. It is interesting to
note that aryldisulfides (e.g.; 2,2'-dithiopyridine)
induce SR Ca^{2+} release and exhibit potent biphasic
influence on [^{3}H]ryanodine binding sites, activation
at $\leq$ 1 μM; inhibition at > 1 μM (Figure 10).

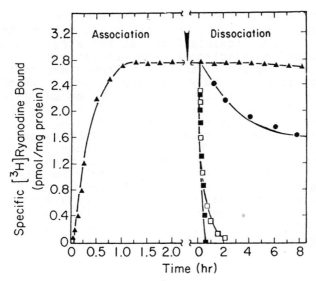

Figure 9 Dissociation of [³H]ryanodine (2.5 nM) from the Ca²⁺-[³H]ryanodine-receptor equilibrium complex affected by 10 μM unlabelled ryanodine (●), Ag⁺ (■), or DTNB (□). Dissociation is initiated after 2 h equilibration by addition of excess chemical (arrow) and measuring residual receptor occupancy at subsequent times.

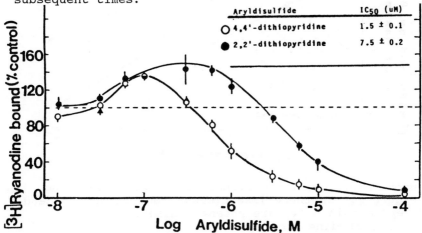

Figure 10 Aryldisulfides have a biphasic response on [³H]ryanodine binding sites suggesting complex kinetics.

Xanthene dyes including rose bengal and erythrosin B
have potent insecticidal activity which reflect their
ability to absorb visible light in the singlet state
which excites an electron to the triplet state (64).
The excited electron of the dye molecule may transfer
its energy to oxygen to yield singlet oxygen which may
serve to oxidize cellular constituents or may perhaps
directly influence the redox state of cellular thiols.
Preliminary data suggest a potent and selective
interaction between xanthene dyes and the ryanodine
receptor (Figure 13). The light-independent toxicity
of rose bengal is nearly 5-fold greater than its
congener erythrosin B (64). Surprisingly rose bengal
shows a 6-fold higher inhibitory potency at the
ryanodine receptor than erythrosin B (Figure 13). The
possible role of redox coupling of these compound with
critical thiols of the ryanodine receptor   is
currently being investigated.

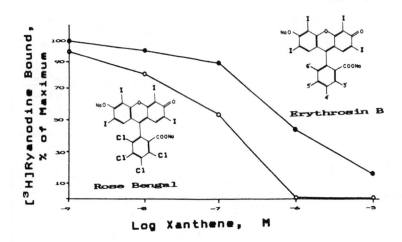

Figure 11   Xanthene insecticides rose bengal (o) and
  erythrosin B (●) are potent inhibitors of
  [³H]ryanodine binding.

VIII. Conclusions

The ryanodine receptor is a key component of excitation-contraction coupling in striated muscle. The highly allosteric nature of the ryanodine receptor channel complex is demonstrated by its susceptibility to modulation to a number of physiologically relevant agents including nucleotides, xanthenes, divalent cations (e.g., Ca^{2+} and Mg^{2+}), and calmodulin. In addition, receptor function and therefore muscle function can be influenced by diverse classes of drugs and toxicants. Characterization of ryanodine receptor structure and function in crustacean and insect muscle is not only of fundamental importance to our understanding of EC coupling in Arthropoda, but will undoubtedly define significant differences relative to the typical vertebrate receptor. Differences at the level of the ryanodine receptor are predicted by the remarkable selectivity exhibited by ryanodol and its didehydro analog in effecting toxicity in arthropods, including Musca domestica, Periplaneta americana, and Procambarus clarkii.

Presently, the complexity of the ryanoid ring system precludes access to synthetic congeners of ryanodol which could potentially serve as selective insecticides. However, biochemical studies exploring the mechanisms by which xenobiotics influence ryanodine binding in conjunction with calcium transport measurements permit rational strategies for the development of insecticides which influence a non-neuronal target.

REFERENCES

1.   Usherwood, P.N.R., Insect Muscle, Academic Press, London, 1975.

2.   Allen, D.G. and Blinks, J.R. Nature (1978) 273, 509-513.

3.   Blinks, J.R., Rudel, R., and Taylor, S. (1978) J. Physiol. London 277, 291-323.

4.   Fabiato, A. (1981) J. Gen. Physiol. 78, 457-497.

5.   Martonosi, A.N. (1982) Transport of calcium by sarcoplasmic reticulum. In "Calcium and Cell Function," Vol. III, (W.Y. Cheung, Ed.), Academic Press, N.Y., p.38-102.

6.   Reeves, J.P. and Sutko, J.L. (1979) Proc. Natl. Acad. Sci., USA 76, 590-594.

7.   Martonosi, A.N. Physiol. Rev. (1984) 64, 1240-1320.

8.   Ford, L.E. and Podolsky, R.J. (1970) Science (Wash. D.C.) 167, 58-69.

9.   Fabiato, A. and Fabiato, F. (1977) Circ. Res. 40, 119-129.

10.   Endo, M. and Nakajima, Y. (1978) Nature New Biol. 246, 216-218.

11.   Miyamoto, H. and Racker, E. (1982) J. Membr. Biol. 66, 193-201.

12.   Meissner, G. (1984) J. Biol. Chem. 259, 2365-2374.

13.   Chamberlain, B.K., Volpe, P., and Fleischer, S. (1984) J. Biol. Chem. 259, 7540-7546.

14.   Smith, J.S., Coronado, R., and Meissner, G. (1985) Nature (Lond.) 316, 446-449 .

15. Vergara, J., Tsien, R.Y., and Delay, H. (1985) Proc. Natl. Acad. Sci., USA 82, 6352-6356.

16. Volpe, P., Salviati, F.D., Virgilio, F.D., and Pozzan, T. (1985) Nature, (Lond.) 316, 347-349.

17. Jenden, D.J., and Fairhurst, A.S. (1969) Pharm. Rev. 21, 1-25.

18. Fairhurst, A.S. and Hasselbach, W. (1970) Eur. J. Biochem. 13, 504-509.

19. Fairhurst, A.S. (1973) Biochem. Pharmacol. 22, 2815-2827.

20. Sutko, J.L., Willerson, J.T., Templeton, G.H., Jones, L.R., and Besch, H.R. Jr. (1979) J. Pharmacol. Exp. Ther. 209, 37-47.

21. Sutko, J.L. and Kenyon, J.L. (1983) J. Gen. Physiol. 82, 385-404.

22. Hilgemann, D.W., Delay, M.J., and Langer, G.A. (1983) Circ. Res. 53, 779-793.

23. Seiler, S., Wegener, A.D., Whang, D.D., Hathaway, D.R.,and Jones, L.R. (1984) J. Biol. Chem. 259, 8550-8557.

24. Fabiato, A. (1985) Fed. Proc. 44, 2970-2976.

25. Meissner, G. (1986) J. Biol. Chem. 261, 6300-6306.

26. Lattanzio, F.A., Schlatterer, R.G., Nicar, M., Campbell, K.P., and Sutko, J.L. (1987) J. Biol. Chem. 262, 2711-2718.

27. Palade, P. (1987) J. Biol. Chem. 262, 6142-6148.

28.   Endo, M., Kakuta, Y., and Kitazawa, T.
       (1981) In "The Regulation of Muscle
       Contraction: Excitation-Contraction
       Coupling"(Grinnell, A.D. and Brazier,
       M.A.B., eds.) Academic Press, N.Y. pp.
       181-195.

29.   Kim, D.H., Ohnishi, S.T., and Ikemoto, N.
       (1983) J. Biol. Chem. 258, 9662-9668.

30.   Kirino, Y., Sakabe, M., Shimizu, H. (1983)
       J. Biochem. 94, 1111-1118.

31.   Nagasaki, K. and Kasai, M. (1984) J.
       Biochem. 96, 1769-1775.

32.   Fabiato, A. (1985) J. Gen. Physiol. 85,
       189-320.

33.   Palade, P. (1987) J. Biol. Chem. 262,
       6135-66141.

34.   Zorzato, F., Salviati, G., Facchinetti,
       T., and Volpe, P. (1985) J. Biol. Chem.
       260, 7349-7355.

35.   Volpe, P., Palade, P., Costello, B.,
       Mitchell, R.E., and Fleischer, S. (1983)
       J.   Biol. Chem. 258, 12434-12442.

36.   Escudero, B. and Gutierrez-Merino, C.
       (1987) Biochim. Biophys. Acta. 902, 374-
          384.

37.   Abramson, J.J., Trimm, J.L., Weden, Lyle,
       and Salama, G. (1983) Proc. Natl.   Acad.
       Sci., USA 80, 1526-1530.

38.   Salama, G. and Abramson, J. (1984) J.
       Biol. Chem. 259, 13363-13369.

39.   Trimm, J.L., Salama, G., and Abramson,
       J.J. (1986) J. Biol. Chem. 261,
       16092-16098.

40.   Pessah, I.N., Stambuk, R.A., and Casida,
       J.E. (1987) Molec. Pharm. 31, 232-238.

41.   Desmedt, J.E. and Hainaut, K. (1977) J. Physiol. (London) 265, 565-585.

42.   Mickelson, J.R., Ross, J.A., Reed, B.K., and Louis, C.F. (1986) Biochem. Biophys. Acta 862, 318-328.

43.   Ohnishi, S.T., Waring, A.J., Fang, S.G., Horiuchi. K., Flick, J.L., Sadanaga, K.K, and Ohnishi, T. (1986) Arch. Biochem. Biophys. 247, 294-301.

44.   Palade, P. (1987) J. Biol. Chem. 262, 6149-6154.

45.   Chamberlain, B.K., Volpe, P., and Fleischer, S. (1984) J. Biol. Chem. 259, 7547-7553.

46.   Somlyo, A.V. (1979) J. Cell Biol. 80, 743-750.

47.   Ferguson, D.G., Schwartz, H.W., and Franzini-Armstrong, C. (1984) J. Cell Biol. 99, 1735-1742.

48.   Franzini-Armstrong, C., and Kenney, L.J., Varriano-Marston, E. (1987) J. Cell Biol. 105, 49-56.

49.   Pessah, I.N., Waterhouse, A.L., and Casida, J.E. (1985) Biochem. Biophys. Res. Commun. 128, 449-456.

50.   Fill, M., and Coronado, R. (1988) Trends Neurosci. 11, 378-401.

51.   Takeshima, H., et al. (1989) Nature 339, 439-445.

52.   Waterhouse, A.L., Holden, I., and Casida, J.E. (1984) J. Chem. Soc., Chem. Commun., 1265

53.   Waterhouse, A.L., Holden, I., and Casida, J.E. (1985) J. Chem. Soc., Perkin Trans. 2, 1011.

54.   Pessah, I.N., Anderson, K.W., and Casida,
           J.E. (1986) Biochem. Biophys. Res. Commun.
           139, 235-243.

55.   Pessah, I.N., Francini, A.O., Scales,
           D.J., Waterhouse, A.L., and Casida, J.E.
           (1986) J. Biol. Chem. 261, 8643-8648.

56.   Waterhouse, A.L., Pessah, I.N., Francini,
           A.O., and Casida, J.E. (1987) J. Med.
           Chem. 30, 710-716.

57.   Abramson, J.J., Buck, E., Salama, G.,
           Casida, J.E., and Pessah, I.N. (1988) J.
           Biol. Chem. 263, 18750-18758.

58.   Ma. J., Knudson, C.M., Campbell, K.P.,and
           Coronado, R. (1989) Biophys. J. 55, 237a.

59.   Abramson, J.J., Trimm, J.L., Weden, Lyle,
           and Salama, G. (1983) Proc. Natl. Acad.
           Sci., USA 80, 1526-1530.

60.   Salama, G. and Abramson, J. (1984) J.
           Biol. Chem. 259, 13363-13369.

61.   Trimm, J.L., Salama, G., and Abramson,
           J.J. (1986) J. Biol. Chem. 261,
           16092-16098.

62.   Bindoli. A. and Fleischer, S. (1983) Arch.
           Biochem. Biophys. 221, 458-466.

63.   Murphy, A.J, (1976) Biochem. 15, 4492-
           4496.

64.   Heitz, J.R. (1982) In Insecticide Mode of
           Action, Coats, J.R. ed., Academic Press,
           New York, 429-4517.